SOUTH SAN FRANCISCO LIBRARY
3 9048 04852129 2

RECYCLING IN AMERICA

A Reference Handbook

Second Edition

Other Titles in ABC-CLIO's

CONTEMPORARY WORLD ISSUES

Series

Abortion, Second Edition	Marie Costa
America's International Trade	E. Willard Miller and Ruby M. Miller
American Homelessness, Second Edition	Mary Ellen Hombs
Animal Rights	Clifford J. Sherry
Childhood Sexual Abuse	Karen L. Kinnear
Children's Rights	Beverly C. Edmonds and William R. Fernekes
Domestic Violence	Margi Laird McCue
Euthanasia	Carolyn Roberts and Martha Gorman
Gangs	Karen L. Kinnear
Gay and Lesbian Rights	David E. Newton
Global Development	William Savitt and Paula Bottorf
Legalized Gambling	William N. Thompson
Militias in America	Neil A. Hamilton
The New Information Revolution	Martin K. Gay
Organized Crime	Patrick J. Ryan
Rape in America	Rob Hall
Sports Ethics	Lawrence H. Berlow
United States Immigration	E. Willard Miller and Ruby M. Miller
Violent Children	Karen L. Kinnear
Wilderness Preservation	Kenneth A. Rosenberg

Books in the Contemporary World Issues series address vital issues in today's society such as terrorism, sexual harassment, homelessness, AIDS, gambling, animal rights, and air pollution. Written by professional writers, scholars, and nonacademic experts, these books are authoritative, clearly written, up-to-date, and objective. They provide a good starting point for research by high school and college students, scholars, and general readers, as well as by legislators, businesspeople, activists, and others.

Each book, carefully organized and easy to use, contains an overview of the subject; a detailed chronology; biographical sketches; facts and data and/or documents and other primary-source material; a directory of organizations and agencies; annotated lists of print and nonprint resources; a glossary; and an index.

Readers of books in the Contemporary World Issues series will find the information they need in order to better understand the social, political, environmental, and economic issues facing the world today.

RECYCLING IN AMERICA

A Reference Handbook

Second Edition

Debra L. Strong

CONTEMPORARY WORLD ISSUES

ABC-CLIO
Santa Barbara, California
Denver, Colorado
Oxford, England

Library of Congress Cataloging-in-Publication Data

Strong, Debra L.
Recycling in America : a reference handbook / Debra L. Strong —
2nd ed.
p. cm. — (Contemporary world issues)
Includes bibliographical references and index.
1. Recycling (Waste, etc.)—United States. 2. Recycling (Waste, etc.)—Law and legislation—United States. 3. Recycling (Waste, etc.)—Information services. I. Title. II. Series.
TD794.5.S77 1997 363.72'82'0973—dc21 97-41346

ISBN 0-87436-889-8 (alk. paper)

03 02 01 00 99 98 10 9 8 7 6 5 4 3 2 (cloth)

ABC-CLIO, Inc.
130 Cremona Drive, P.O. Box 1911
Santa Barbara, California 93116-1911

This book is printed on acid-free ∞ recycled paper (20 percent post-consumer content).

Manufactured in the United States of America

Contents

Figures and Tables

Figures

Tables

Acknowledgments

No book, be it a reference work or a novel, is written by the author alone. For this reason, I would like to gratefully acknowledge the support and special efforts of those who have helped make this book possible, and I'd like to thank God for giving me the strength and ability to finish this project.

During the time that I worked on this second edition of "*Recycling in America*," my life took a great many turns. I met, traveled with, and married my new husband, Richard Strong, and I moved from the mountains to the city. Needless to say, as a result of these events my lifestyle changed markedly. Richard, new to the research/writing process, has been understanding, supportive, helpful, and encouraging to the *n*th degree and beyond. Thank you, Richard, for all your patience and love.

Throughout this challenging time my daughter Tracy, now a teenager, has also encouraged me. She has never let me consider the excuse that I had too much going on in my life to finish this project. And almost every day she asked me, sternly, "Mom, did you work on your book today?" Thanks, Tracy, for hanging in there with me.

Lastly, toward the end of this project, Andrew Barnett Smith, the beloved, 15-year-old son of my dear friend Linda Smith, died in an automobile accident. I would like to dedicate this book to his memory and make an impassioned plea that all who read this remember to *always* wear their seatbelts.

Recycling: A State of Flux

The field of recycling is in constant motion. Although recycling has occurred in nature for billions of years (there is approximately the same amount of water on the planet now as there was when the planet was created), we humans have only adopted recycling on a large scale in the last 30 years or so.

Recycling is, in its simplest sense, taking a used product and remaking it into a new, useful product instead of discarding it as waste. For most of human history, recycling was typically done only when people lacked the resources to manufacture a new product from virgin materials. By and large, recycling was a concept employed by the needy.

Recycling for the environment, as opposed to recycling for sustenance or immediate necessity, is a relatively recent idea. An increased awareness of the limits of natural resources, in conjunction with the population explosion, has led to the conclusion that recycling is an idea whose time has come.

In the United States, the practice of recycling became socially accepted—instead of stigmatized—only recently, and in the past few years, it has become downright fashionable, the "politically correct" thing to do.

An Overview of the Past: Recycling for Necessity or Profit

The problem of managing solid waste dates back to the time when people began to settle in permanent communities. When this evolution from a transient lifestyle to a more sedentary one took place, wastes of all kinds began to accumulate.

Originally, the accumulation of garbage was treated as an aesthetic issue: Garbage looked repulsive, smelled bad, and attracted vermin. Many generations passed before people began to discover that a host of health problems were also associated with their wastes. In time, procedures for collecting and removing wastes from populated areas were established, but little thought was given to the ultimate disposal of solid wastes. It was often assumed that unlimited amounts of space were available in which to dispose of refuse and that wastes could be discarded without long-term ill effects.

Occasional efforts at recycling were made, but usually only in the face of an immediate and pressing need. For example, much recycling was accomplished in the United States during World War II, but it was done as a necessity born from the scarcity of strategic materials. Thus, the nation collected and recycled scrap metals and paper for the war effort, not for the environment, and when the war ended, so did this collective effort to recycle.

In the heady economic climate of the postwar years, consumers wanted new cars, appliances, and other products that made them feel part of a new era. Indeed, everything had to be new, for new meant better, and industry was more than happy to oblige.

The 1950s and early 1960s gave rise to the beginnings of the highly successful disposable industry, an industry that sold the idea that single-use, throwaway items were absolute necessities of a modern lifestyle. Disposables began with paper cups and napkins and ultimately evolved to include diapers, razors, cameras, and even contact lenses. "Use it once and throw it away" became consumers' national motto, and ease and convenience became two of the most desirable qualities in consumer products.

The realization that the concept of "unlimited" might be unfounded with respect to the generation of wastes and the inherent abilities of nature began to dawn on U.S. citizens in the late 1960s. This sudden insight might be traced, in part, to the photographs of Earth from outer space provided by burgeoning space exploration programs. The pictures showed our bright blue planet sitting alone, surrounded by infinite, black space—an undeniably closed and finite system. Such photos afforded a new perspective on planetary limits for scientists and laymen alike.

During this period, we were also discovering that toxins we had discarded into the land, air, and water were having deleterious effects on the environment. With the 1962 publication of Rachel Carson's *Silent Spring*, the idea that our species could so negatively impact life on Earth became a chilling reality. The environmental movement was born, the first Earth Day was organized, and recycling for environmental reasons was undertaken for the first time.

The Environmental Protection Agency (EPA) was created in 1970 as a governmental response to the public's environmental concerns, and its Office of Solid Waste was formed specifically to examine the problems caused by the generation and disposal of wastes. With the passage of the Resource Conservation and Recovery Act (RCRA) in 1976, Americans at last began to look at their waste disposal habits. Among other things, the act mandated that dumps be replaced with regulated and closely monitored landfill facilities, and it was expected that stringent restrictions on waste disposal would encourage recycling on a more widespread basis.

State and local governments were given the primary responsibility for municipal solid waste management, but RCRA also granted the EPA regulatory and assistance powers to make the job easier. However, it still took the better part of another 20 years for recycling to become a mainstream movement instead of the fringe activity of committed environmentalists.

From the early 1970s to the mid-1980s, recycling by individuals and by the businesses that were salvaging used commodities and selling them for reprocessing was occurring on a limited but profitable basis. All parties involved were thriving economically, for there was a relatively small supply of recyclable commodities (such as newspaper, aluminum, and glass) and a steady demand for them. During this period, only select groups were participating in recycling, and most of them were doing so in hopes of making a profit. Helping the environment was a secondary consideration—a bonus point used to sell the idea.

However, a drastic change in the industry occurred in the late 1980s when it became increasingly apparent that cities on both coasts were facing significant problems with the disposal of their trash. Prices for getting rid of garbage were skyrocketing because of the limited amount of land available, and tipping fees (the fee-per-ton to dump trash at a waste disposal facility) at landfills were becoming prohibitive. These costs and the media's intense coverage of the *Mobro 4000*, the infamous wandering "Garbage Barge," began to stir the nation's conscience.

When photographs of medical wastes and the corpses of marine mammals washing up on beaches began to appear in national magazines and on television, a sense of urgency was at long last created with regard to waste disposal. The idea that "there is no such place as away" finally became a reality to most Americans.

Sudden Changes in the Industry

The twentieth anniversary of Earth Day, in 1990, finally seemed to bring home the idea that in addition to finite resources and the toxic threats of pollution, there was also a limit on the amount of space available to deposit wastes. People also became aware that problems such as deforestation, topsoil erosion, groundwater pollution, and global warming were all associated with their wastes. Soon *landfill, nontoxic,* and *biodegradable* became household words, and the concept of a "green lifestyle," referring to an enviromentally responsible way of living, became socially and financially marketable. After using these words casually for a few months, however, Americans found out that they had a great deal more to learn.

Biodegradability has since become a controversial topic. At one time, it was generally believed that if a product was biodegradable, it could be disposed of without consumer guilt, for, in short order, nature would reclaim it; thus, one could throw away copious amounts of trash with impunity. With this in mind, advertisers made *biodegradable* a key word in many successful marketing campaigns during this period. The concept of biodegradability seemed to be valid—until it was applied to plastic bags. Ultimately, researchers found that bags labeled "biodegradable" actually degraded only to a certain point, leaving bits of plastic behind. Furthermore, because they only degraded in the presence of light, they would not decompose when buried in landfills.

Suddenly the whole notion of biodegradability was in the spotlight, and research into what it really meant and how it applied to our solid waste situation was begun. When Dr. William Rathje, from the University of Arizona, started publicizing the results of his studies of deep core samples taken from landfills, one thing became clear: The fact that an item was technically biodegradable did not necessarily mean that it would, in fact, biodegrade in a landfill.

The sanitary landfills of today are not the dumps of the past. Because the waste materials are compacted and covered with soil repeatedly, air, light, and water do not circulate freely among them. Without such circulation, the biological breakdown process involved in biodegradation cannot take place in a normal fashion. Consequently, the decomposition that occurs in landfills is generally limited to the process of drying out, or desiccation.

In his studies, Rathje found easily recognizable chicken drumsticks and hot dogs that had been buried ten years previously—certainly enough time to biodegrade. (He was able to date his finds by checking the newspapers and phone books that had been thrown away at the same time.) Soon, environmentalists began pointing out that even biodegradable materials would remain virtually unchanged in landfills for a very long period of time. And as the true meaning of the word *biodegradable* was exposed, it became apparent that biodegradation was not the answer to the solid waste problems of modern society.

Whether a grocery sack was plastic or paper really didn't matter, for both products used resources, and neither would biodegrade in a landfill. (Today's environmental answer to this dilemma is to take reusable cloth bags to the supermarket and refuse both of the disposable varieties.)

A suddenly enlightened public began demanding accessibility to recycling programs on a widespread basis, believing that if biodegradability was not the answer, then recycling must be. This demand, coupled with the glut produced by the ensuing flood of materials into the commodities' markets, changed the face of the recycling industry almost overnight.

Prior to the late 1980s, recycling in the United States had been accomplished mainly through individual and grassroots efforts. Small businesses had sprung up to successfully handle the limited supplies and demands of the recycling industry. Newspaper and aluminum can drives by churches and scout troops were common, and buy-back centers provided individuals and groups with places to take their recyclables and redeem them for cash. Drop-off centers where items could be left for recycling without payment also gained in popularity and were often located for the convenience of the community. But when the general population began to recycle its newspapers, aluminum cans, and glass en masse, the supply of these recyclables increased dramatically, leading to a significant drop in market prices for those products. Thus, the more people recycled, the less money they received for their efforts.

But even as the economic feasibility of recycling declined, the issue of solid waste continued to rear its ugly head. When Americans were repeatedly forced by the EPA, state and local governments, grassroots environmental activists, and the media to examine the reality of the enormous amounts of waste they were producing, recycling was still regarded as one of the primary solutions.

With the ever-increasing volume of solid waste, it was apparent that more landfills were needed. But where would they be placed? After all, no one wanted a landfill in his or her own neighborhood. Average Americans knew very little about landfills, but what they did know was bad. Landfills had a reputation for being ugly and smelly and for causing pollution, which was all people felt they needed to know. As a result, the NIMBY (Not in My Backyard) syndrome was born.

As the EPA published reports on the current state of solid waste in American communities, it became increasingly apparent that proper solid waste management—or the lack thereof—would determine whether or not the human race would smother under its own garbage. In 1989, with every American producing an average of 3.5 pounds of garbage each day,[1] these wastes had to go somewhere, and there was a limit to what could be recycled. Therefore, landfills were still considered a necessary part of the overall solution.

Integrated Solid Waste Management: A Hierarchy

Although the EPA first published its model of integrated waste management in 1989, most people involved in the waste industry did not become familiar with it until 1990. The hierarchy of waste treatments, originally described in the EPA's *The Solid Waste Dilemma: An Agenda for Action*, encouraged the country to integrate the use of several approaches to solid waste management:

1. Source Reduction

Source reduction, the first approach for any solid waste management program, would ensure that wastes would be reduced at their sources. For example, the use of cloth napkins instead of paper ones would eliminate the need for either disposal or recycling, and a significant quantity of trash would no longer be sent to the landfill.

In response to the concept of source reduction, the packaging industry has increasingly addressed public concerns about solid waste by creating packaging with less volume and the same strength. For example, in the past five years, advances in the resins used to produce plastic bags have reduced the material requirements by up to 30 percent without a loss of bag strength.[2] Also, most grocery stores now boast a wide variety of refills for numerous liquid detergents and other cleaning products, which in turn eliminates a considerable amount of packaging. Similarly, a one-gallon plastic milk container that weighed 120 grams in the 1960s now weighs only 65 grams.[3]

Another success story in the field of source reduction is the expanded use of aluminum foil in the packaging industry. For example, to package 65 pounds of coffee in 2-pound cans requires 20 pounds of steel. But the same 65 pounds of coffee, packaged in a flexible vacuum pack incorporating foil, requires only 3 pounds of packaging. This would mean that even if all of the new packaging ends up in a landfill, the solid waste impact would be reduced by 88 percent compared to the conventional steel coffee can.[4]

2. Recycling

The second step in the waste management hierarchy was considered to be recycling. Recycling would, optimally, remove and recover reusable resources from the waste stream. This step also included the natural recycling process of composting organic wastes such as food scraps and yard trash.

Properly promoted, recycling and source reduction were expected to remove a large percentage of the materials that were being landfilled at that time in the forms of yard waste (17.9 percent), food waste (7.9 percent), paper and paperboard (41 percent), glass (8.2 percent), metals (8.7 percent), and plastics (6.5 percent) (see Figure 1.1).

In reality, by 1994 the amount of solid waste disposed in these categories, after recycling, changed only slightly by percentage: yard waste declined to 14.8 percent, paper and paperboard to 32.9 percent, glass to 6.4 percent, and metals to 6.3 percent. Meanwhile, the amount of discarded plastic actually increased to 9.5 percent, and food waste increased to 8.5 percent (see Figure 1.2).

However, considering the five-year increase in population, and the per capita increase in waste materials (4.4 pounds in 1994 compared to 3.5 pounds in 1989), these figures represent significant

Figure 1.1 Materials Discarded into the Municipal Waste Stream in 1986 (percent of total)

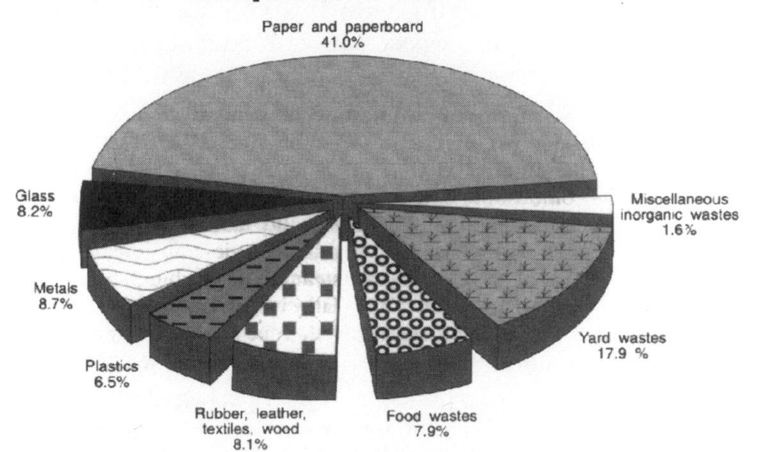

Source: Adapted from U.S. Environmental Protection Agency, *Decision-Maker's Guide to Solid Waste Management,* Washington, DC: U.S. Environmental Protection Agency, 1989.

Figure 1.2 1994 Net Waste Disposal—159.8 Million Tons (after recycling)

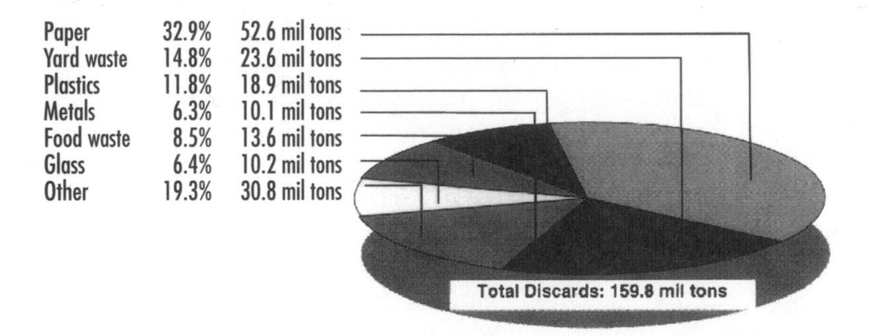

Source: Printed from the MSW Factbook, Ver. 3.0, Office of Solid Waste, U.S. Environmental Protection Agency, Washington, DC..

progress in solid waste management. Also, the amount of solid waste actually recovered from the original amount generated had increased markedly by 1994. Of the 81.3 million tons of paper and paperboard being generated, 35.3 percent was recovered (for re-use or recycling). Over 22 percent of yard waste, 23.4 percent of glass, and 35.9 percent of all metals were also being recovered from the waste stream (see Table 1.1). The efforts of American citizens to recycle and compost were paying off. In view of such progress, the EPA decided that rather than providing a national recycling goal, as it had in the past, individual states should both run their own waste management programs and set their own recycling goals (see Table 1.2).

Table 1.1 Generation and Recovery of Materials in MSW, 1994 (In millions of tons and percent of generation of each material)

	Weight Generated	Weight Recovered	Recovery as a Percent of Generation
Paper and paperboard	81.3	28.7	35.3%
Glass	13.3	3.1	23.4%
Metals			
Ferrous metals	11.5	3.7	32.3%
Aluminum	3.1	1.2	37.6%
Other nonferrous metals	1.2	0.8	66.1%
Total metals	15.8	5.7	35.9%
Plastics	19.8	0.9	4.7%
Rubber and leather	6.4	0.5	7.1%
Textiles	6.6	0.8	11.7%
Wood	14.6	1.4	9.8%
Other materials	3.6	0.8	20.9%
Total materials in products	161.4	41.9	25.9%
Other wastes			
Food wastes	14.1	0.5	3.4%
Yard trimmings	30.6	7.0	22.9%
Miscellaneous inorganic wastes	3.1	Neg.	Neg.
Total other wastes	47.8	7.5	15.7%
TOTAL MUNICIPAL SOLID WASTE	209.2	49.4	23.6%

Includes wastes from residential, commercial, and institutional sources; Neg. = Less than 50,000 tons or 0.05 percent; Numbers in this table have been rounded to the first decimal place.

Source: Adapted from U.S. Environmental Protection Agency, *Characterization of Municipal Solid Waste in the United States: 1995 Update, Executive Summary.* Washington, DC.

Table 1.2 1996 State Recycling Rates Survey Chart

State	MSW Generation (Million tons)	Recycling/ Waste Reduction Goal
Alabama	5.4	25% [1]
Alaska	0.6	10% by 1996
Arizona	Unknown	None
Arkansas	1.9	30% by 1995; 40% by 2000
California	44	25% by 1995; 50% by 2000
Colorado	4.4	50% by 2000
Connecticut	3	40% by 2000
Delaware	0.8	None
District of Columbia	0.15 [2]	45% by1994
Florida	25	30% by 1995 [3]
Georgia	9.6	25% by mid-1996
Hawaii	2	25% by 1995; 50% by 2000
Idaho	Unknown	None
Illinois	14.2	25% by 1996 [4]
Indiana	5.8	35% bu 1996; 50% by mid-2001
Iowa	3.2	59% by mid-2000
Kansas	3	None
Kentucky	4	25% by 1997
Louisiana	N/A	25%
Maine	1.3	50% by 1998
Maryland	N/A	20%
Massachusetts	6.8	46% by 2000
Michigan	13.5	25% by 2005
Minnesota	4.7‡	36% rural; 50% metro by 1996
Mississippi	2.3	25% by 1996
Missouri	7.5	40% by 1998
Montana	0.82‡	25% by 1996
Nebraska	1.7	25% by 1996; 40% by 1999; 50% by 2002
Nevada	3.1	25% by 1994
New Hampshire	1.15	40% by 2000
New Jersey	7.8 MSW‡; 15.9 total	60% by 1995
New Mexico	2‡	25% by 1995; 50% by 2000
New York	26.8‡	50% by 1997
North Carolina	10	40% by mid-2001 [5]
North Dakota	0.5	40% by 2000
Ohio	17.2%‡	50% by 2000 (25% MSW/50% industrial)
Oklahoma	3	None
Oregon	3.4‡	50% by 2000
Pennsylvania	9.5	25% by 1997

Table 1.2 (Continued)

State	MSW Generation (Million tons)	Recycling/ Waste Reduction Goal
Rhode Island	N/A	70%
South Carolina	3.8	30% by 1997
South Dakota	0.45	25% by mid-1996; 50% by mid-2001
Tennessee	7.3‡	25% by 1995
Texas	25‡	40% [6]
Utah	1.9	None
Vermont	0.5	40% by 2000
Virginia	8	25% by 1997
Washington	N/A	50%
West Virginia	1.8	20% by 1994; 30% by 2000
Wisconsin	5.4‡	None
Wyoming	0.5	None [7]

[1] Alabama has no deadline.
[2] District of Columbia includes residential waste only.
[3] Florida counties with more than 50,000 residents.
[4] Illinois counties with more than 100,000 residents.
[5] North Carolina waste reduction goal.
[6] Texas waste reduction goal under consideration.
[7] Wyoming plans to set a voluntary goal this year.
‡ 1994 figures.

Source: Adapted from *Waste Age Magazine,* Aug. 1996. © 1996 by Environmental Industry Associations.

3. Waste Combustion

The next step in the waste management hierarchy was waste combustion, or incineration, with energy recovery. Waste-to-energy (WTE) facilities could take refuse, burn it at extremely high temperatures to eliminate toxic emissions, and use the heat thus produced to generate steam to run turbines, helping to meet a city's electric needs. The relatively small amount of ash left over from the burning process could then be safely landfilled.

At this time, approximately 120 such WTE plants are operating in the United States. About eight states manage over 70 percent of the capacity of WTE fuel (see Table 1.3).

Some environmental groups view incineration with a jaundiced eye, for unless extremely high combustion temperatures are

Table 1.3 Waste-to-Energy Plants

State	Number	Capacity (Tons/day)
Florida	13	18,400
New York	12	11,600
Massachusetts	8	10,200
Pennsylvania	6	8,600
Minnesota	15	6,700
Connecticut	6	6,500
Virginia	6	6,300
New Jersey	6	6,100

Source: Adapted from I. Winston Porter, Ph.D, *Trash Facts III*. Leesburg, VA: Waste Policy Center, 1996.

maintained, toxins are released into the air. The EPA, however, stands by its model and permitting regulations. (A Japanese firm has come up with a novel solution to this problem: A large, neon temperature gauge is placed at the top of a stack, enabling the public to read the temperature from miles away and report any potential problems.)

4. Landfilling

Although landfilling is the last step in the integrated waste management plan, it *is*, nonetheless, a part of it. There are approximately 6,000 landfills currently operating in the United States, almost half of which belong to local governments. Another 15 percent are run by private companies, and the remainder belong to other solid waste or governmental authorities.

The EPA's view is that because landfills will always be necessary for some waste, it is important to make the process of landfilling as safe as possible.

Subtitle D

Following the introduction of the integrated waste management hierarchy, new and stricter guidelines for the siting, maintenance, monitoring, and closure of sanitary landfills were published by the EPA in the *Federal Register* in 1991; they are known as Subtitle D of the Resource Conservation and Recovery Act of 1976 (RCRA).

The intent of Subtitle D is not only to protect the environment and ensure the public's health by closing unsafe landfills but also to encourage the public to utilize methods other than landfilling for the disposal of solid wastes. At the time Subtitle D was published, the EPA estimated that it would result in the closing of up to 50 percent of the nation's landfills between 1992 and 1997. In fact, according to EPA estimates in 1996, approximately 36 percent, or 3,581 landfills, have closed since 1992.

Another result of Subtitle D, accurately predicted at its time of inception, has been the steady increase in costs for existing and new landfills built in compliance with the regulation. Higher disposal costs at landfills have in turn contributed to higher recovery rates in recycling programs in many areas of the country, as well as boosting the economic incentives for the waste industry's promotion of source reduction and safe waste-to-energy systems.

The Present: A Continuing State of Evolution

Public education programs initiated in the early 1990s to teach the importance of recycling from an environmental standpoint proved to be quite successful. As a result, people have generally become less concerned with the money they can earn from their trash. Spurred on by their success, environmental educators are continuing to focus on the idea of recycling for a cleaner, greener Earth, with more trees, purer water, and fresher air.

What started the recycling movement was cash incentives, e.g., trading cans for money; now the emphasis has changed to recycling for the good of the planet. In fact, educational campaigns have been so successful that people are now demanding to recycle whether or not they receive any cash back.

During the 1990s, the United States witnessed record-breaking levels of recycling with regard to products traditionally deemed recyclable. And beyond that, as the concept of recycling for the environment gained in popularity, people started insisting that additional products, such as plastic bags and containers, steel "tin" cans, and office paper, be made recyclable. In 1997, many more items are widely accepted in recycling programs than were accepted in 1992.

However, the industry's economic climate has always been one of extreme volatility, and this continues to be the case. In the late 1980s, the combination of the steady decline of market prices and the increase in services required in the new era of environmental consciousness resulted in the demise of many recycling businesses that lacked the capital to survive such major economic changes. Increased competition from trash haulers, who are uniquely suited to run large, commercial recycling programs, sounded the death knell for many grassroots recycling projects. In 1991, recycling in the United States reached an all-time high, but in most parts of the country it had also moved into the global corporate arena.

The huge waste hauling industry had evolved over generations of trash production, stimulated by the need to handle society's sanitary disposal requirements in an efficient and cost-effective manner. In the past, the industry was primarily concerned with removing wastes, not recycling them. But when the demand for recycling services became as common as the demand for rubbish removal, the waste haulers took advantage of their expertise in collecting wastes and quickly learned how to give the public what it wanted: convenient recycling with the least amount of effort.

Curbside Recycling

Consumers had demanded that recycling be made easy. Sorting and collecting piles of newspapers, bottles, and cans and then transporting them to the local buy-back or drop-off center took too much time for a frenetically paced, time-conscious society. The recycling industry began to respond by offering new programs, and the concept of curbside recycling was refined.

Attempts at curbside recycling, in which recyclables were picked up from individual homes and businesses, had been started in the past by smaller recyclers, but the tremendous costs of vehicle fleet maintenance and fuel, coupled with the labor-intensive nature of the business, proved prohibitive to most.

Several different methods of curbside recycling are still being evaluated by both the public and the industry. In some programs, recyclables such as newspapers, glass, and aluminum beverage cans are sorted into three or more containers and left on the curb on a weekly or biweekly basis.

Variations of this program allow homeowners to commingle, or mix, certain products within the three containers. For instance, glass and plastics or aluminum and steel "tin" cans may be placed together. Drivers of the recycling trucks may sort the plastic from the glass at truckside, depositing aluminum and steel cans together to later be mechanically separated with a magnet at the processing center.

In some curbside programs, all cleaned recyclable items are commingled in a single container. This version of curbside recycling has become increasingly popular in the late 1990s for three major reasons: (1) it's quicker and easier for the hauler to pick up the materials; (2) technology has created more efficient sorting machines to deal with commingled recyclables; and (3) it's more popular with the average consumer who wants to spend as little time as possible on recycling chores.

Although some other programs allow the commingling of trash and recyclables, effectively releasing the consumer from any recycling responsibilities, the programs are often unsuccessful because the commingling of trash and recyclables generally results in a poor-quality product that no buyer is willing to purchase.

In a market in which they are flooded with products, buyers of recyclables can be highly selective about the quality of the materials they accept. In the recent past, for example, glass collected for recycling was accepted or rejected solely on the basis of its cleanliness, i.e., clean glass was typically accepted and paid for, but dirty glass was rejected and often ended up in a landfill. However, in the mid-1990s the standards were raised, and now glass must not only be clean but it must also be sorted by color (clear, brown, and green) in order to be acceptable to most glass buyers in the United States.

Product Contamination

The concept of contamination in recycling is becoming increasingly familiar to the consumer. Contamination refers to the presence of undesirable materials mixed in with the desired recyclable products. Glass buyers, for example, will consider plastic a contaminant in any load they look at, just as plastic buyers consider glass a contaminant in their loads.

In general, all buyers of recyclables have very specific definitions of and limits on contaminants. Entire truckloads of recyclables often have to be landfilled because of extensive contamination. Therefore, the recyclables collector/hauler must either educate the public on this issue or pay higher labor costs to sort and clean materials when the consumer refuses to do so.

Most recycling companies provide lists defining acceptable and unacceptable materials in their individual programs. When consumers are unsure about the acceptability of a specific item, the general rule of the industry is, "When in doubt, leave it out!"

Different programs have different specifications, depending on the markets available for the recyclable products in a given locale.

Materials Recovery Facilities

As the recycling industry has grown, so have the places that handle recyclables. In the past, haulers and the general public usually deposited their trash in one central location, known as a transfer station. The garbage was further compacted at this site and subsequently transported to a sanitary landfill.

From this transfer station concept came the materials recovery facility, or MRF (pronounced "murf"), where recyclable materials can be brought to be sorted, compacted, and shipped to buyers for reprocessing. MRFs range in structure and function from simple to grandiose. The simplest are merely large warehousing areas where recyclables are brought to be hand-sorted for further transfer. More complicated versions feature mechanical separators designed to sort selected products (the magnetic separator, for example, combines a conveyor belt and a powerful magnet to separate aluminum cans from steel cans). State-of-the-art MRFs include machines that can take any variety of commingled recyclables and sort them mechanically with a variety of blowers, shakers, and weight-sensitive devices.

Most MRFs include at least one products baler. A baler is a machine that, when fed a continuous amount of a specific recyclable product, such as newspaper, compacts it to create uniform bales that are generally held together by wires, much like hay bales. Balers can vary in size and in the type of bale they produce, and some are designed for specific products.

The Economics of Recycling

As the recycling business has evolved into a major industry, it has also been forced to deal with the economics of big business. Many small recyclers have failed because they could not deal with the financial realities stemming from glutted materials markets and increased consumer demand for services. But regardless of the size of a business, the current economics of recycling remain the same: The dollar cost of recycling far outweighs the dollar value of the recyclables. It may take up to five times the amount of money a recyclable product is worth to collect, process, and transport it to a buyer.[5]

There have been short periods, over the last few years, when prices rose to record high levels, spurring on new growth in the industry. But the cycle always seems to return to the lower end because consumers can inevitably produce more waste than the industry can keep up with. One example of this volatility has been the market for used corrugated cardboard. In the spring of 1995, prices for this type of cardboard reached $200 per ton, but by winter of 1996 the prices were hovering in the $20 to $30 range.

In other cases, it seems as though part of the industry is harming itself, especially in the case of the packaging plastic PET (polyethylene terephthalate), most commonly used as soft drink bottles. Where once this plastic was the "star" of plastics recycling, prices have plummeted and markets are disappearing because chemical companies are selling virgin PET resin at cut-rate prices. (Virgin resin normally costs 30 cents to 45 cents per pound but was being sold at 18 cents to 25 cents per pound in the winter of 1996. It costs 25 cents to 35 cents per pound to convert postconsumer PET bottles into flakes or pellets.)

Thus, the bottom line is that the recycling industry today, at least from the collection point of view, is driven by consumer demand, not by profit. It is difficult to run a successful business when one is trying sell products whose values are far from stable, in spite of the fact that overhead costs (fuel, insurance, personnel) remain largely unchanged. The profits for waste haulers may not exist at this time, but people are still insisting that they be allowed to recycle because "it's the environmental thing to do."

It should also be noted that several studies have shown that the higher the recycling rate in a city, the more cost effective its recycling program becomes.[6] As a result, although most trash-hauling businesses are motivated to promote recycling in order to retain customers, the reward for their success may end up being a profit.

There are some areas of the recycling industry where economic advantages have already been realized, and this is most evident in the growing number of businesses and technologies based on using recycled materials in the products they manufacture. Where one may have been hard-pressed to find products containing significant amounts of recycled materials five or ten years ago, it is hardly a problem in today's market. Even McDonald's Big Mac™ wrappers bear the imprint, "Made with 33% recycled paper (15% postconsumer content)." And a plastics-molding company in Iowa, Hettinga Equipment, Inc., has created and is further developing the technologies necessary to take all sorts of mixed plastic "garbage" and turn it into usable products, such as marine plywood, while making a profit at the same time. As the economics slowly start to turn around in some sectors, this will in turn promote the potential for profit in other areas.

Environmental economists also point out that in what has been termed a "sustainable" economic system (one based on the real costs to the environment resulting from the production and transportation of goods and materials), recycling would be financially cost effective from all aspects. In such a system, the prices of products made from virgin materials would be prohibitive, encouraging manufacturers to use recycled materials instead. However, under such a system, the concept of curbside recycling would become even less cost effective than it is now, due to increased transportation and energy costs.

Legislation is being considered in various parts of the country to place taxes on virgin materials and/or on the products made from them in order to stimulate the recycling economy, but no such laws have been passed at this time. However, several states, including California, Colorado, and New York, do have tax incentive laws based on promoting research and development in the recycling industry (see Chapter 5).

Pay-as-You-Throw

Another cost-effective way to manage waste and increase recycling rates was developed in the late 1980s, when some innovative solid waste management specialists and trash haulers combined the fervor for recycling with a lesson on source reduction by offering "pay-as-you-throw," volume-based trash fees.

In such programs, a homeowner pays less money for throwing out less trash. In order to produce less trash, recycling, composting, and source reduction efforts are generally necessary, and in this way, economic incentives for recycling as well as source reduction have been created. The concept gives consumers the same economic perks that they receive for practicing water or energy conservation in their homes: lower utility bills each month.

There are two basic styles of pay-as-you-throw systems. In one, households are billed for each bag or container of trash that gets picked up by the waste hauler. In the other, which is less common, residents pay bills based on the weight of their trash. According to the EPA, there are now over 2,000 communities across the United States who have implemented pay-as-you-throw programs, and

Figure 1.3 Recycling Symbol

Source: American Paper Institute

officials have reported average waste reductions ranging from 25 to 45 percent.

The only potential downside to pay-as-you-throw is the possibility of illegal dumping, where people who do not wish to pay more for the amount of trash they create take their refuse to any clandestine place they can find and dump it. However, communities that have instituted this system have generally not found this to be the case, especially when they provide other ways to reduce trash legally, such as easy access to recycling programs and composting programs for yard wastes.

Completing the Cycle

In analyzing the economics of recycling, it is important to understand what recycling really means. Although the general public has typically focused on the collection of recyclables, this is, in fact, only a single part of the cycle of recycling.

The three chasing arrows of the recycling symbol (see Figure 1.3) represent the three parts of the recycling cycle:

- Collection is the first part of the cycle, in which materials are gathered together by consumers, recycling centers, and curbside collection programs and taken to a MRF.
- In reprocessing, the second part of the cycle, the recyclables are bought back by manufacturers and remade into new products.
- Purchasing the products made from recycled materials is the final part of the cycle.

Although the collection part of the cycle has been well understood by the public for ten years or more, the other two components have only recently become common knowledge, thanks to the more thorough educational programs taught in the schools and presented in the media. But the fundamental cause of the economic dilemma that faces much of the recycling industry today still remains in many sectors: Too many people believed for too long that they were recycling if they sorted their trash and took items to the local recycling center or had them picked up.

In reality, these people were only participating in the very first step of the cycle. And though industrial interests handle the bulk of the second step—the reprocessing—the primary responsibility for completing the last portion of the cycle rests, again, with the consumer. For until the demand for products made from recycled materials catches up to or exceeds the supply of collected recyclables, the economic stability of the industry will remain at risk.

This failure to complete the cycle and the resulting financial hardships help to explain the growing role played by waste haulers in the field of recycling. Hauling corporations, which have other assets to carry them through the leaner times, are most likely to survive until the widespread completion of the cycle creates a healthier financial picture for recycling.

Hopes for the Future

In the meantime, industrial, governmental, educational, and environmental interests are banding together to educate the public more thoroughly about solid waste issues and the concept of integrated solid waste management. At the same time, efforts to develop markets for recyclable materials are growing by leaps and bounds, with many new consortiums springing up seemingly overnight. Most states now include these efforts within their economic development programs, which tie into their own supplies-purchasing practices.

On the educational front, priorities include teaching the 3Rs—reduce, reuse, and recycle—to the general populace. Successful implementation of these concepts is key to the first two approaches of the integrated waste management hierarchy (source reduction and recycling), which will, in turn, lead to the conservation of both resources and landfill space:

- *Reduce* refers to source reduction, specifically targeting unnecessary packaging and disposable single-use products.
- *Reuse,* a concept that fits closely with reduce, refers to using items over and over instead of just one time (e.g., using cloth napkins and sponges instead of paper napkins and paper towels).
- *Recycle* is now seen as only part of the total solution to solid waste problems, and the emphasis on completing the cycle has become a new priority, along with composting, both at home and at the municipal level.

Some states have helped the educational process along by passing legislation mandating certain types of recycling, banning the disposal of some materials, or even prohibiting the sale of certain packaging materials (see Chapter 5).

But stimulating old markets for recyclables while at the same time creating new ones is the main focus of the recycling industry today. One way to accomplish this goal is through legislation requiring taxes on virgin materials or products made from them. Much of the current recycling legislation encourages price preferences for goods made from recycled materials, but none has gone so far yet as to tax virgin materials. Although some tax incentives have been offered for efforts to research, develop, and establish recycling technologies, the main tool of the industry has been marketing—selling the idea of "green consumerism" to the public.

"If you're not recycling, you're throwing it all away" and "If you're not buying recycled, you're not really recycling" are two prevalent slogans in this campaign. More and more businesses are cropping up to cater to the new "green" consumers by offering only "environmentally friendly" products, i.e., those that are nontoxic, recyclable, made from recycled materials, and/or designed to help consumers recycle.

Additionally, new products are constantly being created for the "green" businesses: cat litter made from recycled drywall gypsum, animal bedding made from shredded newspapers, recycled plastic lumber for construction, floor tiles containing recycled glass, carpeting made from recycled plastic soft drink bottles, and carpet padding made from recycled tires. A new clothing fabric, Tencel™, made from recycled wood fibers, feels more like silk. Attractive buttons are being made from recycled magazines, and Huffy, one of America's oldest bicycle manufacturers, has created a mountain bike line whose frames are made from 25 percent recycled steel, with other components made from recycled plastics. Lately it seems that new products are being developed at an ever-increasing rate, and where once these products were only marketed to a select group of eco-conscious citizens, the general public is now the target group.

The more consumers purchase these products and others that carry a "made from recycled materials" label, the more successful the recycling industry will become.

A New Focus: Source Reduction

It seems, however, that most "new" ideas on how to best deal with our wastes keep returning to the basic model of the integrated waste management hierarchy.

As the end of the twentieth century approaches, it is becoming apparent that a new focus has emerged from the field of solid waste management. Whereas most waste haulers used to concentrate their efforts on the collection side of their business when dealing with recycling, they are now changing over to a more active promotion of source reduction.

Many haulers are involved in coordinating programs that encourage and facilitate the composting of yard wastes. Some are even creating joint ventures with lawn mower manufacturers, such as Toro™, to further the use of mulching mowers that pulverize grass clippings, altogether eliminating the need for bags full of grass to be disposed of at landfills.

New literature, published for their customers, indicates a serious effort on the part of waste haulers to truly address the problem of solid waste at its source: the consumer. Just as the collection of recyclables and the buying of products made from recyclable materials has become a more accepted part of mainstream societal values in the late 1990s, so it seems that now, following the educational efforts at source reduction, it has become fashionable to carry lunch boxes, use cloth napkins, refuse additional bags in department stores, and use personalized ceramic or plastic cups, instead of disposable ones, for coffee in the morning.

But it's not just the waste haulers who have gotten on the source reduction bandwagon—source reduction has also increasingly become the top priority in federal, state, and local solid waste management planning. Most solid waste planning offices currently focus more of their programs on reducing the amount of materials used in the workplace than on recycling. The EPA's WasteWi$e program, for example, works specifically with the business community on this issue, recognizing that 40 percent of all the municipal solid waste generated in the United States comes from this source. Founded in 1994, this program now boasts the participation of over 430 companies.

As part of this public focus on the first two steps of the waste management hierarchy, large-scale composting programs are also becoming more popular and are receiving more funding for research and development than ever before. Since yard and food wastes make up approximately 30 percent of the waste stream in this country, composting has always seemed to be a logical path to take to accomplish waste reduction goals, but in the past it seemed too difficult and messy a process to encourage on a wide-scale basis. Times have changed, however, and almost every state planning office now offers information on backyard composting and/or citywide yard waste composting programs.

New Technologies: Businesses Born of the Need to Recycle

An encouraging sign of the times is the new emphasis placed on inducing manufacturers to make their products in ways that facilitate recycling. Thus, automakers around the world are beginning to create programs for labeling plastic car parts with universal resin identification codes, enabling the final salvager to recycle as many pieces as possible.

The automotive industry has also begun recycling its own waste right in its factories. Some carmakers in Detroit have purchased equipment that enables them to take scraps of plastic right from the floor, place it into a molding machine, and create new components from materials that used to be hauled away to the landfill. The automotive industry is also focusing on source reduction by making cars from more lightweight materials but that have the same structural strengths as their heavier and more voluminous predecessors.

Similarly, large appliance makers are manufacturing their goods so that they are ultimately easier to disassemble into recyclable scrap, which has resulted in a higher rate of recycling for these items. In 1995, 41.8 million appliances were recycled at a rate of 74.8 percent, which is a respectable increase from only 40 percent at the beginning of this decade.[7]

Today's more environmentally educated public and sharper industries have also stimulated and supported a new form of research called life-cycle analysis. This research is meant to show the true environmental cost of any product, from cradle to grave. In other words, the costs of energy, resources, and labor are considered from creation through disposal (or recycling), and companies are using such data to demonstrate that one product is better or more environmentally sound than another. This field has proven to be fairly effective in supporting arguments for sustainable economics, where the cost of a product or service is ultimately determined by its effect on the environment.

How to Use This Book

Wherever there are recycling programs, there are changes. As with any reference book, numerous changes are bound to occur between the time it is written and the time of publication. Thus, this book is meant, first of all, as a guide to the fundamentals of recycling and, secondly, as an extensive collection of resources where the reader can find further, up-to-the-minute information on everything that has anything to do with recycling in the United States.

This guide is divided into eight chapters and provides the reader with the following:

1. An introductory, overall look at the way solid waste issues evolved in our country and how recycling fits into the solutions
2. A chronology of recycling, beginning in ancient times and ending with a focus on the last 25 years or so of the recycling industry's history in the United States
3. An overview of some of the prominent people in America who have affected the evolution of recycling
4. An examination of the most commonly recycled commodities, together with interesting facts and figures on these goods, a look at how each product is manufactured from virgin materials, and a description of how each product is actually recycled, noting the additional goods that can, in turn, be made from it
5. An overview of recycling and solid waste legislation in the 50 states
6. An extensive directory of public and private agencies and

organizations involved in this field, on both the state and national levels

7. An annotated list of references, including various print resources on recycling (in addition to those that are specific to locales and, as such, may be listed in Chapter 5 under a state agency or organization), as well as a guide to available nonprint resources, such as videotapes that deal with recycling and solid waste issues
8. A directory of e-mail addresses and Internet home pages that can provide additional, up-to-the-minute solid waste and recycling information both in the United States and in Canada.

Finally, a comprehensive glossary of terms used in the various branches of the recycling industry is included.

It is my hope that any person seeking information on recycling in the United States will be able to find what is needed—or a source in which to locate it—through this book.

Notes

1. U.S. Environmental Protection Agency. *The Solid Waste Dilemma: An Agenda for Action.* Washington, DC: U.S. Environmental Protection Agency, 1989.
2. Plastic Bag Information Clearinghouse. *The Life of a Plastic Bag.* Pittsburgh, PA: Plastic Bag Information Clearinghouse, 1993.
3. U.S. Environmental Protection Agency. MSW Factbook, Ver. 3.0. Washington, DC: U.S. Environmental Protection Agency, 1996.
4. The Aluminum Association. *Aluminum: The 21st Century Metal.* Washington, DC: The Aluminum Association, 1994.
5. Miller, Chaz. *The Cost of Recycling at the Curb.* Washington, DC: National Solid Wastes Management Association, 1993.
6. Morris, David. *Recycling and the New York Times.* Washington, DC: Institute for Local Self-Reliance, 1996.
7. American Iron and Steel Institute. Internet: http://www.steel.org, 1997.

Chronology 2

The history of recycling is inextricably tied to the history of solid waste. Therefore, a partial history of garbage is presented in this chronology, along with the origins of many currently recycled products.

12000 B.C. Egyptians use the first glass, in the form of beads.

10000 B.C. People begin to establish permanent settlements. For the first time, garbage begins to accumulate.

7000 B.C. Man first learns to mold glass into usable shapes.

1500 B.C. The first jars and bottles are made out of glass.

400 B.C. The first municipal dump in the Western world is established in ancient Athens. Virtually anything considered unwanted waste is left in the dump.

20 B.C. Glassblowing is discovered as a techniques for making containers.

A.D. 105 Ts'ai Lun invents paper in China, using reclaimed materials such as rags, worn fishnets, hemp, and China grass, solving a pressing shortage of printing materials at a time when the only materials for printing were made from pure silk. To reward his efforts, the Emperor Ho-Ti makes Ts'ai Lun a marquis.

200 The Romans create the first sanitation force: teams of two men walking along the city streets, picking up garbage, and throwing it into a wagon.

1388 The English Parliament bans waste disposal in ditches and public waterways.

1400 The waste from Paris is piled so high outside the city gates that it interferes with the city's defenses.

1415 A dump makes history as it is heroically "captured" during a Portuguese attack on a Moroccan city. The attackers thought it was a strategic hill.

1551 The first recorded example of packaging: a German papermaker, Andreas Bernhart, begins placing his paper in a wrapper that bears his name and address.

1608 English settlers build a glass-melting furnace in Jamestown, Virginia, the first in North America.

1690 Paper recycling is born in America when the first paper mill is established by the Rittenhouse family on the banks of the Wissahickon Creek, near Philadelphia. The paper at this mill is made from recycled rags.

1757 Benjamin Franklin starts the first street cleaning program in North America.

1776 The first metal recycling in America occurs when patriots in New York City topple a statue of King George III, melt it down, and make it into 42,088 bullets.

1785 The first cardboard box made in America is manufactured in Philadelphia by Frederick Newman.

1810 The "tin can" is patented in London, England, by Peter Durand.

1834 A law is enacted in Charleston, West Virginia, that protects vultures from hunters because of the important role that they play in the city's garbage removal program.

1854 Frenchman Henri-Étienne Sainte-Claire Deville found a way to produce aluminum through a chemical process, naming the ore "bauxite."

1858 The Mason jar is invented, allowing fruits and vegetables to be preserved.

1865 An estimated 10,000 hogs roam the streets of New York City, gorging on garbage.

1868 John Wesley Hyatt creates the first plastic from nitrocellulose and camphor; he calls it "celluloid." (By the mid-1980s, the volume of plastics used in the United States exceeds that of steel.)

1880 Private scavenging companies and municipal crews begin working together to clean up New York. They remove 15,000 horse carcasses from the city streets (city horses have rough lives pulling street cars; their average life expectancy is only two years).

1884 The Washington Monument's top is capped by a small aluminum pyramid (aluminum was then more precious than gold).

1885 The first garbage incinerator in the United States is built on Governor's Island in New York Harbor.

1886 By an incredible coincidence, two men (American Charles Martin Hall and Frenchman Paul L. T. Heroult) simultaneously discover the electrolytic process for producing aluminum that is still used today. It is called the Hall-Heroult process.

1895 King C. Gillette, a traveling salesman, tires of sharpening his razor and creates the disposable razor blade.

1897 New York City's rubbish now gets delivered to a "picking yard." Here it is separated into five grades of paper, three grades of carpet, and four grades of metal. Bagging, twine, rubber, and horsehair are also "picked" out for reuse.

1903 The Owens Automatic Bottle Machine is invented; it is now possible to mass-produce jars and bottles of uniform height, weight, and capacity. Glassblowing ceases to be a major method of production.

1904 Large-scale aluminum recycling begins in Chicago and Cleveland.

The first "junk mail" is delivered when Postmaster Henry Clay Payne authorizes permit mail so that 2,000 or more pieces of third- or fourth-class mail can be posted without stamps by paying a single fee.

1908 The first paper cups are used in vending machines that dispense ice water (a penny a cup).

1912 Plastic packaging is first encouraged when the clear plastic "cellophane" is invented by Swiss chemist Dr. Jacques Brandenberger.

1933 New Jersey obtains a court order stopping New York City from dumping its garbage in the Atlantic Ocean. The Supreme Court upholds the decision a year later but states that the law applies to municipal waste only—commercial and industrial waste is not excluded from ocean dumping.

1935 The first beer can is produced by Kreuger's Cream Ale in Richmond, Virginia, and company sales increase 550 percent over the next six months because it is so convenient not to have to return bottles. (This first beverage can weighed three ounces, compared to today's aluminum version that weighs only one-half ounce, an 83 percent decrease in weight.)

1939 Not to be outdone, the bottling industry fights back with the first "no-deposit, no return" bottles, which are marketed in the guise of "Wisconsin Select" beer.

1943 The aerosol can is invented by two researchers at the U.S. Department of Agriculture.

1944 Dow Chemical invents polystyrene foam and dubs it Styrofoam™.

1948 Fresh Kills Landfill opens in Staten Island, New York. It is destined to become the largest city dump in the world. (The Great Wall of China and Fresh Kills are the only two man-made objects on Earth that are visible from outer space.)

1961 Procter & Gamble begins test-marketing the disposable diaper.

1965 The federal government realizes that garbage has become a major problem and enacts the Solid Waste Disposal Act (SWDA), which calls for the nation to find better ways of dealing with trash. Research is authorized and state grants are provided.

1968 The U.S. aluminum industry begins recycling discarded aluminum products, from beverage cans to window blinds.

1970 On 22 April the first "Earth Day" awakens America to the environmental crises at hand and first introduces the concept of recycling to the general public.

The U.S. Environmental Protection Agency (EPA) is established.

The Resource Recovery Act amends the SWDA, requiring the federal government to publish waste disposal guidelines.

1971 The state of Oregon is the first to create legislation mandating deposits on beverage containers (five cents per container).

1972 The first buy-back centers for purchasing recyclables from the public are opened in Washington State, accepting only beer bottles, aluminum cans, and newspapers.

1973 A bottle made from PET (polyethylene terephthalate) is patented by chemist Nathaniel Wyeth (brother of Andrew Wyeth, the American painter).

1974 The first municipal-wide use of curbside recycling containers occurs in University City, Missouri, where city officials design and distribute a container for collecting newspapers.

1975 The disposable razor is produced and marketed by Gillette.

1976 Newspaper recycling hits its first snag when Vanderbilt University in Nashville, Tennessee, halts its program after newsprint prices drop from $45 to $3 per ton.

Marvin L. Whisman, John W. Goetzinger, and Faye O. Cotton of Bartlesville, Oklahoma, file U.S. Patent 4,073,720 for a method of purifying and reclaiming used lubricating oils.

The Resource Conservation and Recovery Act (RCRA) is passed, mandating the replacement of all dumps with sanitary landfills, which are defined by the EPA as "land disposal facilities that will have no adverse effects on health or the environment." The enforcement of RCRA standards will increase the costs of landfill disposal, which the EPA's Office of Solid Waste believes will make resource-conserving options, such as recycling, more appealing.

1977 The first PET bottle is recycled.

1979 The EPA publishes landfill regulations that prohibit open dumping.

1986 The city of San Francisco meets its goal of recycling 25 percent of the city's commercial and residential waste stream.

Rhode Island becomes the nation's first state to pass a mandatory recycling law for aluminum and steel cans, glass, PET and high-density polyethylene (HDPE) plastics, and newspapers.

1987 The infamous "Garbage Barge," the *Mobro 4000,* sails down the East Coast, through the Bahamas, to Belize and Mexico. The barge is refused permission to dock at each port. After 6,000 miles of sailing, the barge's load of trash is incinerated and the ash buried on New York's Long Island, where the garbage originated.

1988 Several manufacturers of plastic foam–type containers, many of them major petrochemical companies, form the National Polystyrene Recycling Corporation with the objective of recycling 25 percent of foam plastic wastes by 1995.

The Plastic Bottle Institute develops a material identification code system for plastic bottle manufacturers that will help identify bottles suitable for recycling.

New Jersey implements legislation requiring every community to collect at curbside three preseparated categories of recyclables.

1989 Procter & Gamble starts marketing its Spic and Span cleaner in bottles made of recycled PET plastic.

The first two polystyrene recycling plants in the United States open in Massachusetts and New York.

Laws requiring recycling to be an integral part of waste management have been enacted by 26 states.

Seven states mandate separation of recyclables at curbside.

1990 McDonald's announces plans to discontinue polystyrene packaging of its fast food due to consumer protests and the failure of its polystyrene recycling concept. McDonald's and the Environmental Defense Fund later agree on 41 additional steps to cut waste up to 80 percent.

On 4 December, within 22 minutes of each other, both the Coca-Cola Company and PepsiCo, Inc., the two largest soft drink bottlers in the world, make the same announcement: In 1991, they will begin using a recycled PET bottle (made of about 25 percent recycled plastic resin).

1993 The Environmental Defense Fund joins with six of America's largest paper buyers to examine paper use and purchasing practices. The goal of this task force is to build demand for environmentally preferable paper, such as paper with a higher postconsumer recycled content.

1995 According to the Can Manufacturer's Institute, nearly seven out of ten aluminum beverage cans are recycled (approximately 62.7 billion cans).

Sources

Can Manufacturer's Institute. Internet: http://www.cancentral.com, 1996.

EarthWorks Group. *The Recycler's Handbook.* Berkeley, CA: EarthWorks Press, 1990.

Ecenbarger, William. "How Long Can We Keep Dumping on Ourselves?" *The Philadelphia Inquirer Sunday Magazine* (1 January 1989): 8.

Glass Packaging Institute. Internet: http://www.gpi.org, 1996.

Glass Packaging Institute. "Some Historical Notes on Glass." Glass Packaging Institute, no date.

Glenn, Jim. "Containers at Curbside." *BioCycle* (March 1988): 26.

Kane, Joseph Nathan. *Famous First Facts: A Record of First Happenings, Discoveries, and Inventions in American History.* 4th ed. New York: H.W. Wilson, 1981.

McMahon, Jim. Personal interview, 28 December 1991.

National Association for Plastic Container Recovery. *PET Fun Facts.* Internet: http://www.napcor.com, 1997.

National Solid Wastes Management Association. *Garbage Then and Now.* Washington, DC: National Solid Wastes Management Association, 1989.

Rathje, William L. *"The History of Garbage."* *GARBAGE* 5 (September/October 1990): 32–41.

U.S. Environmental Protection Agency Region VIII library, recycling file. 999 18th Street, Denver, CO.

Biographical Sketches

Compiling a section on noteworthy people in the field of recycling is not a simple task. Many women and men have labored for years to promote recycling, largely without recognition. They are the grassroots heroes, striving to develop and implement recycling programs within their home communities.

Many individuals are nationally known in the recycling business, but few people in the industry agree on who the leaders are because they all work together in an admirable fashion to further the common cause of recycling.

The biographies that follow feature exceptional people who have truly made a difference in the national recycling arena, but be assured that they are not the only ones who deserve recognition. It is also important to remember that almost every community has its own VIP in recycling—a person who is often the most valuable resource at the local level.

William M. Ferretti

William M. Ferretti is executive director of the National Recycling Coalition (NRC), a nonprofit organization "committed to promoting recycling as an integral part of waste and resource management and economic development." The NRC has 4,700 members, including environmental and recycling organizations, businesses of all sizes, state and local governments, and individuals, and Dr. Ferretti oversees the activities of the entire coalition.

Before accepting his position at the NRC in 1996, Dr. Ferretti was director of the New York State Office of Recycling Market Development (ORMD). While under Ferretti's direction, the ORMD received national acclaim and recognition for its leadership in the field of recycling and was awarded the first Presidential Award for Sustainable Development in 1996. The ORMD is part of New York State's Department of Economic Development and is the country's oldest public organization devoted to encouraging market development in the realm of recycling.

Dr. Ferretti received his Ph.D. in Resource Management and Policy from the State University of New York and Syracuse University in 1984.

Peter L. Grogan (1949–)

Peter Grogan presently serves as the director of materials recovery and is a partner with the environmental consulting firm R. W. Beck and Associates. Prior to joining the firm, Grogan served for 11 years as the director of a community recycling program in Colorado.

Over the years, Grogan has worked for a number of state governments to develop their recycling/waste reduction programs, including Alabama, California, Indiana, Maine, Massachusetts, and Washington. He has also worked with the Virgin Islands and all of the American flag territories in the Pacific on similar solid waste projects.

The local governments he has assisted with recycling programs include Sitka, Alaska; San Francisco, California; Denver, Colorado; Miami and Orlando, Florida; and Honolulu, Hawaii, as well as Vancouver, British Columbia, and American Samoa.

In addition to his work with the public sector, Grogan has also provided services to develop recyclable commodity markets for private industry. He has worked with U.S. and Japanese paper mills, the plastics industry, and the international banking community and has implemented recycling systems for airports, corporations, U.S. Navy ships, and major theme parks. Currently, Grogan is assisting several municipalities with their plans to build and/or utilize materials recovery facilities.

A member of the editorial board of *BioCycle* magazine, Grogan also writes a monthly column, "Recycling View," for that publication. In addition, he serves on the National Recycling Coalition's board of directors and is a member of its recycling advisory council.

Don Kneass (1945–)

Don Kneass is presently the western regional director for the National Association for Plastic Container Recovery (NAPCOR).

Kneass began his career in recycling in 1971 with the Portland Recycling Team, a nonprofit drop-site recycling business. After moving to Seattle, he established the Washington State Recycling Information Office and its recycling hotline in 1973.

In 1974, Kneass opened Seattle Recycling, Inc., a multimaterial buy-back recycling business. Ten years later, he sold this business to take a position with the City of Seattle Solid Waste Utility as a senior planner responsible for recycling program development. As Seattle's recycling coordinator, his principal responsibility was the development of the city's curbside recycling program.

In February 1988, Kneass began working for Waste Management, Inc., as its Northwest Region Recycling Manager, planning and implementing recycling programs in Alaska, western Canada, the Northwest, and the Rocky Mountain states. His responsibilities have encompassed residential curbside collection, commercial collection, recycling processing facilities and equipment, yard waste collection and composting, tire collection and shredding, and material marketing.

In June 1994, Kneass assumed his current position with NAPCOR, helping communities improve their collection of PET plastic containers. His responsibilities include providing communications, marketing, and technical assistance.

Kneass is presently a member of the Clean Washington Center Policy Advisory Board, a board member of Colorado Recycles, and a member of the King County (Washington) Commission for Marketing Recyclable Materials.

He is a past chairman of the board of the National Recycling Coalition, and a past president of the Washington State Recycling Association. He was also chair of the Washington State Department of Trade and Economic Development's Committee for Recycling Markets.

Kneass also established and served as chief executive officer for the Environmental Enhancement Group in Seattle, Washington, and Los Angeles, California, from 1985 to 1988. This consulting firm provided planning and project management services to government agencies, corporations, nonprofit organizations, and small businesses.

In 1994, he was named Recycler of the Year by the Washington State Recycling Association.

Chaz Miller (1947–)

Chaz Miller is the senior manager of recycling and waste services programs for the National Solid Wastes Management Association (NSWMA) in Washington, D.C. He is also a contributing editor for *Recycling Times* magazine as well as an editorial advisor for *Waste Age* magazine. Currently, Miller writes both an op/ed column and an Internet column, "Recycling the Net," for *Recycling Times*. He has also written a Waste Product Profile series, now in its second edition, which is composed of fact sheets about solid waste and various recyclable commodities.

A nationally known expert on recycling issues, Miller has spoken at recycling and solid waste conferences throughout the United States and Canada, and is often a keynote speaker. As a result of his new Internet column, he is also speaking at Internet workshops held at state and national recycling conferences.

Miller was a member of the board of directors of the National Recycling Coalition for seven years and served as its chairman in 1989. In 1990, he was chairman of the NRC's Markets Development Committee, and from 1988 until 1991, he was also director of recycling for the Glass Packaging Institute (GPI).

With GPI, Miller was responsible for promoting glass recycling activities throughout the United States and also advised members on assisting recycling programs within the frameworks of different states' regulations. He coordinated the actions of state lobbyists and participated in federal legislative sessions, industry conferences, and standards-setting activities.

Before joining GPI, Miller spent 11 years with the Environmental Protection Agency's Office of Solid Waste, working on source separation, resource recovery, hazardous wastes, and information management. He was responsible for several recycling publications and grants, as well as one of the EPA's computerized data-reporting systems.

Miller has testified at congressional and state hearings on recycling issues, and he has supervised several studies on the cost of collecting and processing recyclables. He is also a member of the Montgomery County, Maryland, Solid Waste Advisory Committee.

Gary Michael Petersen (1947–)

Gary Michael Petersen is the founder of five environment-related firms, including Ecolo-Haul Recycling Services, a Los Angeles–based recycling services/environmental education firm established in 1972. He is also the founder of the California Resource Recovery Association and has served on the advisory board of the hazardous and toxic waste certification program of the University of California at Los Angeles (UCLA).

Since 1972, Petersen has worked on numerous recycling and environmental education programs designed to educate both consumers and key opinion leaders about the immediacy of solid waste problems in industrialized countries. In the United States, he has been a consultant for the EPA and the National Science Foundation's Committee for National Recycling Policy, as well as a board member and adviser to the National Recycling Coalition.

Internationally, Petersen has been a consultant to the governments of Great Britain, the People's Republic of China, the Republic of Palau and islands of the Korae State in Micronesia, the provincial governments of Canada, and private industries in various countries.

In 1973, Petersen founded the magazine *Resource News* and served as editor and publisher for a time. He has also produced educational programs on recycling for primary and secondary schools in the Los Angeles area.

Politically active, Petersen lobbied for the California legislation known as SB 650 (1978), which provided grant funds for communities to establish recycling and recycling education programs. In addition, he was responsible for the initial design of AB 2020 (1987), which provided for redemption of glass, aluminum, and plastic containers.

Over the years, Petersen has received numerous awards and letters of recognition for his contributions to the fields of recycling and environmental education. Some of these have come from former President Ronald Reagan, the California Waste Management Board, former California Governor Edmund G. Brown, Jr., the Cousteau Society, and the California Resource Recovery Association. He is also listed in the *International Biographical Reference* (Cambridge, England), *Who's Who in the World*, and *Who's Who in America.*

Since 1988, Petersen has served on the board of the Hazardous Materials Management Program at the University of California. He also designed the curriculum and was an instructor for the municipal solid waste management certificate program at UCLA.

More recently, Petersen was a member of the President's Task Force for the Greening of the White House, which ended in 1994. Currently, he is working on a similar project, the Greening of the United Nations, which began in 1996. Both programs center around making the respective buildings more environmentally sound through the use of sustainable development methods such as using recycled materials and recycling systems, the implementation of water and energy conservation technologies, and the use of source reduction practices.

In May 1997, Petersen finished up a project for Waste Management, Inc. (WMI), after reaching the goal of making recycling more available to the general public, both in the United States and throughout the other parts of the world where WMI operates. When his contract was finished, Petersen had helped the company institute recycling programs for over 8 million homeowners and a half-million businesses throughout the United States, as well as beginning RecycleCanada and similar programs throughout Europe.

Jerry Powell (1946–)

Jerry Powell is the founder and editor-in-chief of Resource Recycling, Inc. As such, he is responsible for the editorial elements of three monthly periodicals: *Resource Recycling* magazine, *Plastics Recycling Update*, and *Container Recycling Report*. Since the 1970s, Powell's expertise has made him a valued and sought-after speaker, enabling him to address more than 150 recycling and solid waste conferences in over 30 states and Canadian provinces, as well as in Hong Kong, Australia, and New Zealand.

Powell is the former president of Resource Conservation Consultants, a recycling firm that he sold in 1988. While working in that capacity he assisted more than 200 state and local governments, as well as recycling businesses throughout North America. Prior to starting the recycling publishing and consulting businesses in 1978, Powell ran Portland Recycling, a community recycling company in Portland, Oregon, for over eight years.

Throughout his career in recycling, Powell has been involved in numerous state and national recycling organizations. He is the founder and a past three-term chair of the Association of Oregon Recyclers, and he has also served as a chair of the board of the National Recycling Coalition. Currently, he serves as chair of the Oregon Recycling Markets Development Council.

Nancy VandenBerg (1946–)

Nancy VandenBerg is the principal of Markets for Recycled Products, a consulting practice she established in 1985 to focus on the importance of purchasing decisions in solid waste issues. VandenBerg pioneered formal recycled product procurement plans and introduced the corporate and nonprofit communities to buy-recycled strategies. Her more recent activities include consultation on green building strategies and construction techniques, as well as waste prevention methods.

Over the years, she has provided critical research on products manufactured from recycled materials. Her feasibility studies on recycled construction products for the Environmental Protection Agency resulted in an insulation guideline published in 1989, as well as four of the five new guidelines developed by the EPA in 1991 and 1992. She has established minimum recycled content standards for the state of Florida as well as the EPA and has published recommendations for numerous recycled products now designated by the EPA for preferential purchasing.

In 1993, while working with other consultants, VandenBerg produced the first evaluation of a procurement-for-waste-prevention program, and also designed recycled content specifications for Tucson, Arizona. In 1996, she and her subcontractors, Susan Kinsella & Associates and Carla Lallatin & Associates, developed *Resourceful Purchasing: A Hands-On Manual* for the Alameda County, California, Source Reduction and Recycling Board. The manual is a hands-on reference tool for buyers that includes both waste prevention and buy-recycled strategies.

VandenBerg's seminars for public and private officials, which were sponsored by the plastics industry in the early 1990s, led to increased acceptance of recycled products. The buy-recycled programs established by many state governments, the EPA, and private corporations are based on concepts she developed.

She has served on the boards of *The Recycled Products Guide* and the *Environmental Building News* and holds founding membership with the Manhattan Solid Waste Advisory Board. She has served on the recycled paper and recycled plastic committees of the American Society of Testing and Materials (ASTM).

VandenBerg drafted and successfully defended critical sections of the 1990 ASTM D5033 *Standard Guide for the Development of*

Standards Relating to the Proper Use of Recycled Plastic, which has led to specification revisions throughout the plastics industry. As of 1997, her recommendations are still leading to revisions in ASTM standards.

Her publications include: "Recycled Materials Procurement II," *Resource Recycling* (November/December 1986); "Working Together to Buy Recycled," *Waste Age* (January 1989); "Plastic Film Recycling," *BioCycle* (December 1991); and "Recycling Policy: What Will It Be?" *PIMA Magazine* (January 1992).

Jeanne Lynn Wirka (1963–)

Jeanne Wirka is a nationally recognized expert on source reduction and recycling, especially as they relate to consumer packaging issues. She is currently the director of the Recycling Economic Development Project at the Californians Against Waste Foundation (CAWF). In this position, she focuses on coordinating the development of local markets for recycled materials with the goals of community economic development in the nine counties of the San Francisco Bay area.

From 1987 until 1991, prior to taking her position with CAWF, Wirka served as the founder and coordinator of the Solid Waste Alternatives Project at the Environmental Action Foundation (EAF) in Washington, D.C. The project was a national campaign aimed at reducing solid waste by promoting changes in product manufacturing and materials use. While Wirka ran this campaign, she also coordinated a coalition of state activists working on uniform solid waste legislation; testified before Congress, the Federal Trade Commission, and state and local legislatures; conducted workshops for policymakers; organized press conferences; and conducted research on the role of materials in solid waste management.

From 1989 to 1991, Wirka served as a member of the board of directors of the National Recycling Coalition. From 1989 to 1990, she was also on both the board of directors and the steering committee of the Source Reduction Council of the Coalition of Northeast Governors (CONEG).

Currently, she sits on the advisory committees of the Environmental Protection Agency's project on life-cycle analysis methodology development and the Office of Technology Assessment's green product design project.

While directing EAF, Wirka wrote and edited its quarterly newsletter and public education materials. She is the author of a nationally acclaimed report on the environmental impacts of plastics packaging, entitled "Wrapped in Plastics: The Environmental Case for Reducing Plastics Packaging." Wirka's other publications include "Life Cycle Assessments: Science or PR?" Solid Waste Action Paper #1 (Environmental Action Foundation, 1991); "Degradable Plastics: The Wrong Answer to the Right Question," with Richard Denison, Ph.D. (Environmental Action Foundation and Environmental Defense Fund, 1989); "The Degradable Plastics Hoax," *Environmental Action* (November/December 1989); and "Plastics Recycling: Missing the Forest for the Plastic Lumber," *Resource Recycling* (December 1989).

Among Wirka's conference papers are: "Plastics Recycling: A Tale of Three Planets" (presented at the Conference on Solid Waste Management and Materials Policy, sponsored by the New York State Legislative Commission on Solid Waste Management, in New York City in 1991) and "The Rise and Fall of Integrated Solid Waste Management" (presented at Recycling Earth's Resources, sponsored by the National Recycling Coalition, in San Diego in 1990).

Facts and Data 4

This chapter provides useful facts and figures about widely recycled commodities, together with an explanation of how these commodities are made from virgin materials, collected for recycling, recycled, and remanufactured. Brief information on several less common recyclables, as well as some odder commodities, has also been included.

Aluminum Cans

Aluminum Can Facts

- Every minute of every day, an average of 119,242 aluminum cans are recycled.
- In 1995, 2 billion pounds of aluminum cans were diverted from landfills.
- More than 50 percent of a new aluminum can is made from recycled aluminum.
- Recycling 1 aluminum can saves the equivalent of enough energy to run a television set for three hours.
- Recycling 1 aluminum can saves enough energy to make 19 more.
- In 1994, recycling aluminum cans saved the energy equivalent of 20.6 million barrels of oil or 12.3 billion kilowatt hours of electricity.

- If aircraft carriers were made from aluminum cans, more than 14 carriers could have been built in 1995 from the 1,008,915 tons of cans recycled.
- Recycling aluminum scrap saves 95 percent of the energy that would have been required to make new aluminum from bauxite ore.
- When aluminum can recycling began in 1972, approximately 15.4 percent of America's cans were recycled.
- Since 1972, approximately 722.8 billion aluminum cans, or 15 million tons, have been recycled. If you placed this many cans end to end, they would reach to the moon some 229 times.
- Aluminum now accounts for 99 percent of the beverage can market; in 1972, the share was about 20 percent.
- In 1989, Americans recycled 49.5 billion all-aluminum beverage cans, bringing the national aluminum can recycling rate to 60.8 percent.
- In 1995, Americans recycled 62.7 billion aluminum cans.
- In the 1980s, aluminum can recycling diverted 12 billion pounds of metal from the U.S. solid waste stream (more than 320 billion cans).
- Today's aluminum beverage can uses 30 percent less metal than the 1972 version.
- In 1972, approximately 21.75 cans could be made from one pound of aluminum.
- In 1996, one pound of aluminum can make approximately 31.073 cans.
- Today's aluminum beverage can weighs only one-half ounce—that's an 83 percent decrease in weight since 1935, when the first beverage can weighed three ounces.
- Aluminum can be recycled and reused indefinitely without a decline in material performance or quality.
- Aluminum can recycling is a closed-loop process, and used cans can be recycled and back on store shelves as new beverage containers in as little as 60 days.
- Every aluminum beverage can has an average of 51.3 percent postconsumer recycled content (a far higher average than any other beverage container).
- It takes 8,760 pounds of bauxite and 1,020 pounds of petroleum coke to produce 2,000 pounds of aluminum.
- When recycled aluminum is used instead of bauxite and petroleum coke, the amount of raw materials needed drops by 95 percent.

- Recycling aluminum reduces associated air pollution by 95 percent.
- Approximately 40 percent of the aluminum used in the United States goes into packaging.
- In 1995, Americans recycled 62.2 percent of their aluminum beverage cans.
- In 1989, 2.9 million tons of scrap aluminum were recovered by the Institute of Scrap Recycling Industries, Inc.

Making and Recycling Aluminum Cans

Although aluminum is an element, it is always found in combination with other elements in its natural state. To make virgin aluminum, bauxite, a reddish, clay-like ore rich in aluminum compounds, must be surface mined; to extract the aluminum, the bauxite is refined to eliminate impurities such as iron oxide. Leading suppliers of bauxite include Brazil, Australia, Ghana, Guinea, Jamaica, and China.

A fine white powder called alumina, a form of aluminum oxide, is produced by the refining process. The alumina is sent to a reduction plant, or smelter, where it is deposited into cells known as pots, which are about 20 feet long, 6 feet wide, and 3 feet deep. A strong, continuous electrical current is passed through the pots, separating the alumina and producing molten aluminum metal, which settles to the bottom. All of these steps, which together involve enormous time and energy, are eliminated by aluminum recycling.

Most aluminum recycled by consumers is in the form of aluminum cans, and aluminum can recycling rates have risen steadily since the 1970s (see Figures 4.1 and 4.2). To prepare an aluminum can for collection, the consumer merely needs to rinse it in order to remove food or beverage remnants. Crushing the can is generally optional; the main advantage is that crushed cans take up less room and can therefore be stored and transported more easily.

Aluminum cans are typically collected from consumers at recycling centers, grocery stores, and through reverse vending machines that give money back for cans deposited. There are also some direct consumer-to-processor sales programs. For example, Reynolds Aluminum, ALCOA, and Anheuser-Busch all buy cans directly from consumers as well as from recycling collection companies.

After the cans are collected, they are physically or mechanically inspected for contaminants, such as glass and cigarette butts.

Figure 4.1 Aluminum Recycling Rates

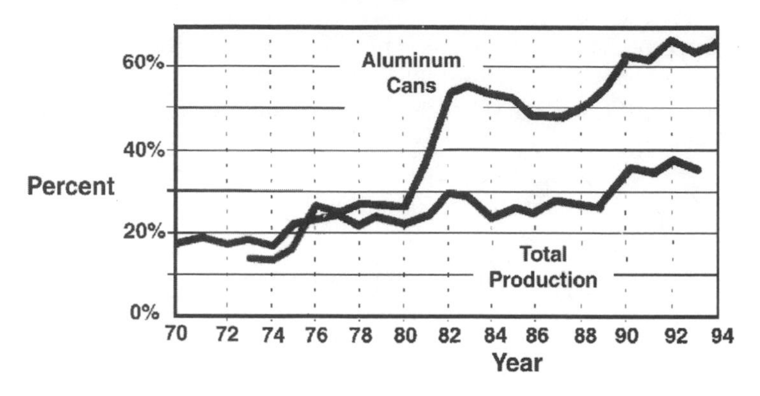

Source: Aluminum Assoc. Aluminum Statistical Review. Printed from the MSW Factbook, Ver. 3.0, Office of Solid Waste, U.S. Environmental Protection Agency, Washington, DC..

Figure 4.2 Aluminum Beverage Cans

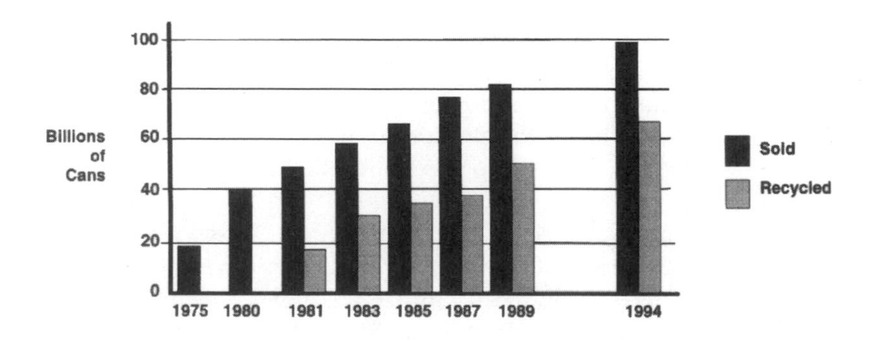

Source: Printed from the MSW Factbook, Ver. 3.0,, Office of Solid Waste, U.S. Environmental Protection Agency, Washington, DC..

After the contaminants have been removed, the cans are usually crushed and packed into bales by a baling machine. These tightly packed bales are then sent to the processing plant to be melted down for reuse.

At most processing plants, the baled cans are shredded to rid them of excess moisture that could cause an explosion in the furnace. The shredded metal is then poured into a huge "pot-line" furnace and heated (the melting point of aluminum is approximately 1,280 degrees Fahrenheit). After the cans have become molten metal, the paint from the outside of the cans rises to the surface and is skimmed off.

The molten aluminum is transferred to another furnace where small amounts of other metals may be mixed with the aluminum to produce the desired characteristics in the finished product (for example, magnesium adds corrosion resistance, and copper adds strength). This newly alloyed molten metal is generally cast into ingots, or large bricks, that can weigh as much as 60,000 pounds and provide enough metal to make more than 1.86 billion beverage cans.

If the aluminum is to be made back into beverage cans, the ingots are shipped to a sheet plant, where they are reheated and rolled into thin sheets of aluminum. These sheets are sent on to a can plant, which transforms the sheet aluminum back into food or beverage cans.

Steel "Tin" Cans

Steel Can Facts

- One ton of steel cans contains 3.8 pounds of tin.
- In 1989, steel and bimetal cans (that is, cans made with steel bodies and aluminum lids) were being recycled at a rate of 21.6 percent nationwide.
- In 1995, the recycling rate of steel and bimetal cans rose to 55.9 percent.
- The average person in the United States uses 12 steel cans each month. Of those 12 cans, the average person throws away 5 and recycles 7.
- 97 percent of all steel cans are used for food packaging.
- About 566 steel cans are recycled every second.
- The average American uses 142 steel cans each year.
- Recycling steel cans saves 74 percent of the energy that would be used to produce them from virgin materials.
- Every day, Americans use more than 100 million steel cans.
- In 1994, 40 billion steel cans, weighing 2,930,000 tons, were used in the United States.

Making and Recycling Steel Cans

Steel is a form of commercial iron characterized by its malleability and by the fact that it has a lower carbon content than cast iron, which is relatively brittle. The easiest way for a consumer to distinguish a steel can from an aluminum one is to apply a magnet. Aluminum is nonferrous (without iron) and therefore nonmagnetic. Steel, on the other hand, is a ferrous metal (containing iron) and will attract a magnet.

Many people are more familiar with steel cans than they are with aluminum ones. The steel can is often called the "tin" can, a misnomer. Actually, there is very little tin in these cans; the base metal is steel, and what little tin is used forms a thin coating on the inside of the can to stabilize the flavors of its food or beverage contents. This coating also enables the manufacturer to sterilize the food by cooking it right in the can.

According to the Steel Recycling Institute (SRI), more than 90 percent of food containers, including the large #10 cans used by food service operations, are made of steel. And steel is America's most recycled material.

Steel cans are prepared and collected for recycling in a number of ways. Most collectors merely request that the cans be rinsed free of food particles (using leftover dishwater, so as not to waste water); others insist that they be crushed.

Some recycling programs collect steel cans separately from other materials, but other programs allow them to be mixed with a variety of recyclables. In fact, they are among the easiest items to remove from a mixture of recyclables because of their iron content: A large magnet is used, often in conveyor form, to separate steel cans from other materials.

Once collected, the steel cans are taken to a processor to be prepared for recycling. The cans may be baled, flattened, or shredded depending on the requirements of the end-market user. Many prepared cans are then shipped to a detinning company, where they are shredded to facilitate the removal of food and paper contamination and then melted down. Various processes are used to remove the tin from the steel, and the detinned steel is sent to a steel mill to be made into a new product. (One ton of steel cans contains 3.8 pounds of tin.) The tin is sold to tin users and recycled in a variety of other products, some of which include more cans.

Some steel mills will purchase steel cans directly, in addition to what they purchase from detinners, to incorporate a certain percentage of tin into their scrap mix. Bimetal cans can be processed in the same manner, and steel aerosol and paint cans can also be recycled. It is best to consult local recycling organizations for their specific preparation guidelines for these containers.

New products made from recycled steel include automobiles, major appliances, and new food and beverage cans. In fact, according to the SRI, all steel packages contain about 28 percent recycled steel because, as has always been the case in the steel industry, recycled steel is a necessary part of the process of making new steel.

Glass

Glass Facts

- Glass is 100 percent recyclable in a true closed-loop system; there are no waste by-products.
- Approximately 41 billion jars and bottles, or 10.3 million tons of glass containers, are manufactured annually in the United States.
- Approximately 650,000 tons of additional glass containers are imported each year.
- Each person in the United States uses about 85 pounds of glass each year.
- Every glass bottle that is recycled can save enough energy to light a 100-watt lightbulb for four hours.
- By recycling one ton of glass, we save the energy equivalent of nine gallons of fuel oil.
- In 1989, glass bottles were being recycled at the rate of 25 percent nationwide.
- In 1995, the glass recycling rate in the United States topped 37 percent.
- Germany recycled almost 40 percent of its glass in 1989.
- About 30 percent of today's average glass beverage container is made of recycled glass.
- Some 75 percent of the glass used in the United States goes into packaging.
- The average person in the United States can save at least seven pounds of glass each month.
- Each day, Americans recycle approximately 13 million glass bottles and jars.

Making and Recycling Glass

There are many different kinds of glass, which vary widely in their basic chemical composition and physical characteristics. There are two major types of commercial glass: "soda-lime-silica" and "special."

Most general recycling programs handle only soda-lime-silica glass, which includes beverage and food containers exclusively. Special glass, which is not widely recycled, includes glass cookware (e.g., Pyrex® and Visionware®), mirrors, windowpanes, windshields, optical glass, lightbulbs, drinking glasses, ceramics, and lead crystal.

Soda-lime-silica glass is, as the name suggests, made from soda (sodium carbonate), limestone (calcium carbonate), and sand (silicon dioxide). The optimum combination of these ingredients—15 percent soda, 10 percent lime, and 75 percent silica—produces a strong, resilient product with a moderately low melting point. The variable qualities of special glass make it incompatible with soda-lime-silica glass for recycling purposes, and for this reason, most recycling programs do not accept special glass and actually consider it to be a contaminant.

Colored glass is produced by adding different metallic oxides to the mixture. For example, iron oxides create the common green color, and brown or amber glass is created when several colors are mixed together. Because glass manufacturers often want a specific color or no color at all, many recyclers require that the glass they purchase from consumers and processors be sorted according to color (clear or "flint," green, and amber).

As glass recycling has become more prevalent, some processors have stopped accepting any green glass because there is simply too much of it. This has become a major market problem, as well as a public relations nightmare, but the reason for it is simple: supply and demand.

Throughout most of the country, most of the green bottles consumed are either imported from other countries or made in California. As a result, there is much more green glass available than the local cullet (cleaned and crushed glass) markets can absorb for the purpose of recycling. Solutions currently being examined include experimenting with using more green glass cullet in making amber glass, and exporting shipments back to countries that import green glass, especially Canada.

To simplify glass recycling in the future, methods of dipping clear bottles in colored inks and acrylics are being investigated. These dipped bottles would not need to be color separated before being recycled because the glass itself would be clear and the dyes could be removed. Other technologies that the glass industry is investigating to make glass recycling easier include automated color separation and mechanized ceramic sorting.

Glass is one of the easiest commodities to recycle, and yet many people still throw bottles in the trash (see Figure 4.3). The consumer merely needs to rinse out any remnants of food products and remove the plastic or metal top. Labels may be left on the containers as they burn off at the high temperatures in the furnace. A local recycler should be consulted for any color-sorting requirements for a specific area.

After being collected, glass must be made "furnace ready"; it is color sorted, if necessary, and made contaminant free. Contaminants include metal caps and lids, plastic caps, dirt, stones, and special glass. Glass that does not meet a buyer's strict specifications is rejected and taken to a landfill.

A glass plant will often designate the number of times that a specific supplier's glass can be rejected before the plant will stop buying from that supplier. The Coors Brewing Company in Colorado, for example, will cease buying from a supplier after three

Figure 4.3 Glass Products in Municipal Solid Waste, 1994

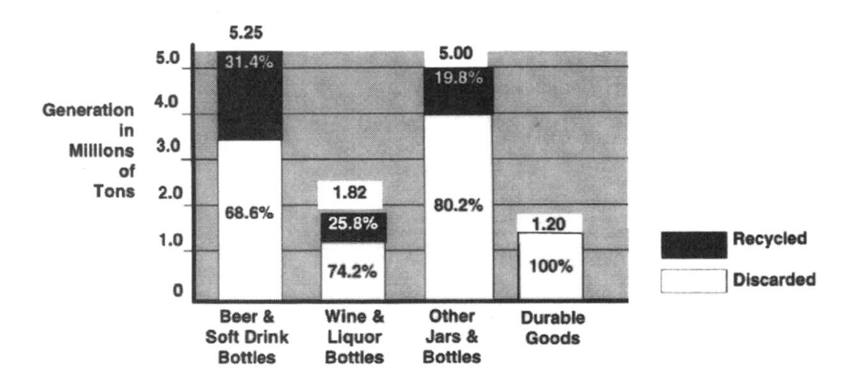

Source: EPA Waste Characterization Report, Frankliln Assoc., 1995. Printed from the MSW Factbook, Ver. 3.0, Office of Solid Waste, U.S. Environmental Protection Agency, Washington, DC..

separate loads have been rejected. Thus, it is very important for the consumer to make sure that the products put into the glass collection bin are clean and uncontaminated. The success of a glass recycling program rests directly with the consumer.

After collection and inspection, the glass is crushed into small pieces called cullet. The cullet is cleaned and then may be mixed directly with whatever additional raw materials are needed to make new glass. This mixture is melted at temperatures of up to 2,800 degrees Fahrenheit. Although the raw materials of glass are still in abundant supply, a great amount of energy is saved by utilizing cullet instead. (For every 1 percent of recycled glass used, there is a 0.5 percent drop in energy costs.)

The melted glass is dropped into a forming machine, where it is either blown or pressed into shape. Newly formed glass containers are then cooled slowly in an annealing lehr, which keeps the glass from becoming brittle.

Most recycled glass is reprocessed back into food and beverage containers, but some goes to other markets. For example, companies that sell stained glass for artistic purposes may use recycled glass, as do some fiberglass companies. Recycled glass is also used by producers of a road composite called glassphalt, which is made of crushed glass and asphalt.

Plastic

Plastic Facts

- Americans throw away 2.5 million plastic bottles every hour.
- 26 plastic soft drink bottles can make one polyester suit.
- Five recycled soft drink bottles will make enough fiberfill for a man's ski jacket.
- Five PET bottles yield enough fiber for an extra-large T-shirt.
- 35 recycled plastic soft drink bottles will make enough fiberfill for a sleeping bag.
- The average household generates 17 pounds of PET bottles annually.
- It takes eight 2-liter soft drink bottles to equal about one pound of PET.
- Five 2-liter PET bottles make one square foot of carpet.
- Half of all polyester carpet manufactured in the United States is made from recycled plastic bottles.
- 1,050 recycled milk jugs can be made into a six-foot park bench.

- Plastics currently comprise approximately 23.9 percent of the volume in U.S. landfills.
- Plastics currently comprise approximately 11.5 percent of the weight in U.S. landfills.
- Of the 14.4 million tons of all types of plastics generated in 1988, only 0.2 million tons (or 1.1 percent) were recovered for recycling.
- In 1995, 18 percent of all plastic bottles and rigid containers were recycled.
- Of all the PET containers produced in 1995, approximately 32 percent of them were recycled (triple the amount in 1990).
- Whenever Americans buy food, they spend $1 out of every $11 on packaging.
- Approximately 30 percent of plastics are used in packaging.
- More than half of the plastics Americans throw away each year, about 6 million tons, are used in packaging.
- The United States makes enough plastic film each year to shrink-wrap the state of Texas.
- Americans use about 5 million tons of plastic wrap each year, very little of which is recycled.
- 40 percent of the plastic trash in the United States consists of plastic bags and film wrappings.
- If only 10 percent of Americans bought products with less plastic packaging only 10 percent of the time, approximately 144 million pounds of plastic could be eliminated from our landfills.
- Each person in the United States uses about 190 pounds of plastic each year.
- More than 5.5 billion 2-liter bottles are sold each year.
- Although recycling keeps about 175 million pounds of PET out of the landfills annually, 535 million pounds of PET are still being thrown away.
- The amount of plastic waste generated has been increasing by about 10 percent per year since 1977.
- Each year, Americans use over 25 billion polystyrene foam cups—enough to encircle our planet 436 times.
- Plastics products have the highest energy values for waste-to-energy incineration facilities (because of their petroleum hydrocarbon content).
- Plastic is virtually immortal: If the Pilgrims had had six-packs, the six-pack rings would still be around today.
- About 100,000 five-gallon HDPE containers are recycled each year by Ben & Jerry's Ice Cream Company.

- Each year, Americans throw away 6 billion plastic-encased disposable pens.
- The average person in the United States throws away two pounds of plastic containers every month.
- Americans throw away over 500 million disposable plastic cigarette lighters each year.
- Every year, the U.S. plastics industry manufactures the equivalent of about ten pounds of plastic for every person on Earth.
- A discarded plastic beverage container has a longer life expectancy than the person who threw it away.

Making and Recycling Plastic

All plastics are made from a basic mixture called resin, which is derived from petroleum oil or natural gas. The resin is sold by chemical companies to manufacturers who remelt the resin, mix in additional chemical additives, and then use pressure molding or extruding processes to make the finished product.

There are over 50 types of plastic resins currently bought and sold in the United States. Fewer than 10 of these are commonly used for packaging, 6 of which were assigned codes by the Society of the Plastics Industry, Inc., in 1988, to facilitate their recycling by consumers (see Figures 4.4 and 4.5).

Figure 4.4 Major Types of Plastics by SPI Codes

SPI CODE	TYPE OF RESIN	EXAMPLE PRODUCTS	% OF PLASTIC
1	PET -Polyethylene terephthalate	Soft drink bottles, medicine containers	0.5%
2	HDPE- High-density polyethylene	Milk and water bottles, detergent bottles, toys	21%
3	PVC - Polyvinyl Chloride	Pipe, meat wrap, cooking oil bottles	6.5%
4	LDPE - Low-density polyethylene	Wrapping films, grocery bags	27%
5	PP-Polypropylene	Syrup bottles, yogurt tubs, diapers	16%
6	PS- Polystyrene	Coffee cups, "clamshells"	16%
7	Other		8.5%

Source: Printed from the MSW Factbook, Ver. 3.0, Office of Solid Waste, U.S. Environmental Protection Agency, Washington, DC..

Figure 4.5 Types of Plastic Packaging

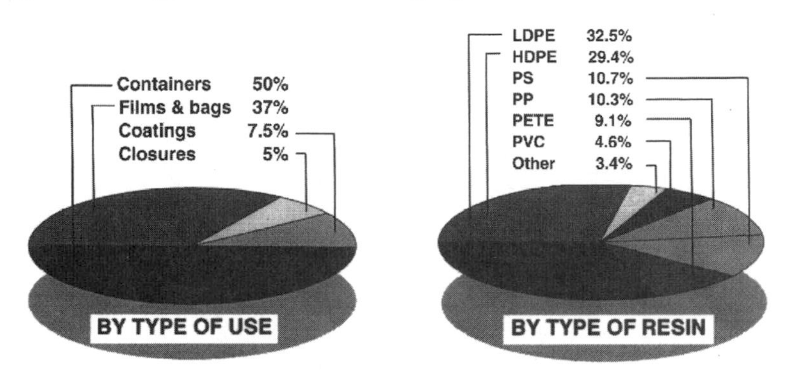

Source: Printed from the MSW Factbook, Ver. 3.0, Office of Solid Waste, U.S. Environmental Protection Agency, Washington, DC..

The industry code, placed on the bottom of the plastic container, consists of a triangle formed by three arrows, with a number in its center and a corresponding letter abbreviation beneath the triangle. The codes range from approximately 0.5 inch to 1 inch in size and can be applied either by molding or imprinting. The codes and their accompanying letters refer to different types of resins (see Figure 4.6).

In general, the basic requirements for recycling a plastic container are the same as those for other containers: The consumer should rinse it thoroughly and remove its top or lid because these are usually made of a different resin that cannot be widely recycled at this time. Recyclers also request that consumers flatten or crush the container as much as possible. This helps the collection program by eliminating a good deal of the dead air space that composes most of the volume in plastic containers. The less space it fills up, the more economical it is for the recycler to store the plastic for later sale.

In some recycling programs, the public is asked to separate two or three resin types. In other programs, certain resin types may be mixed together. However, the majority of plastics recycling programs in this country still accept only the two most commonly recycled resins: PET and HDPE. This is generally believed to be because the process of recycling plastics is still evolving and because these two resin types are the most prevalent in our packaging container industry.

Figure 4.6 Plastic Container Code System

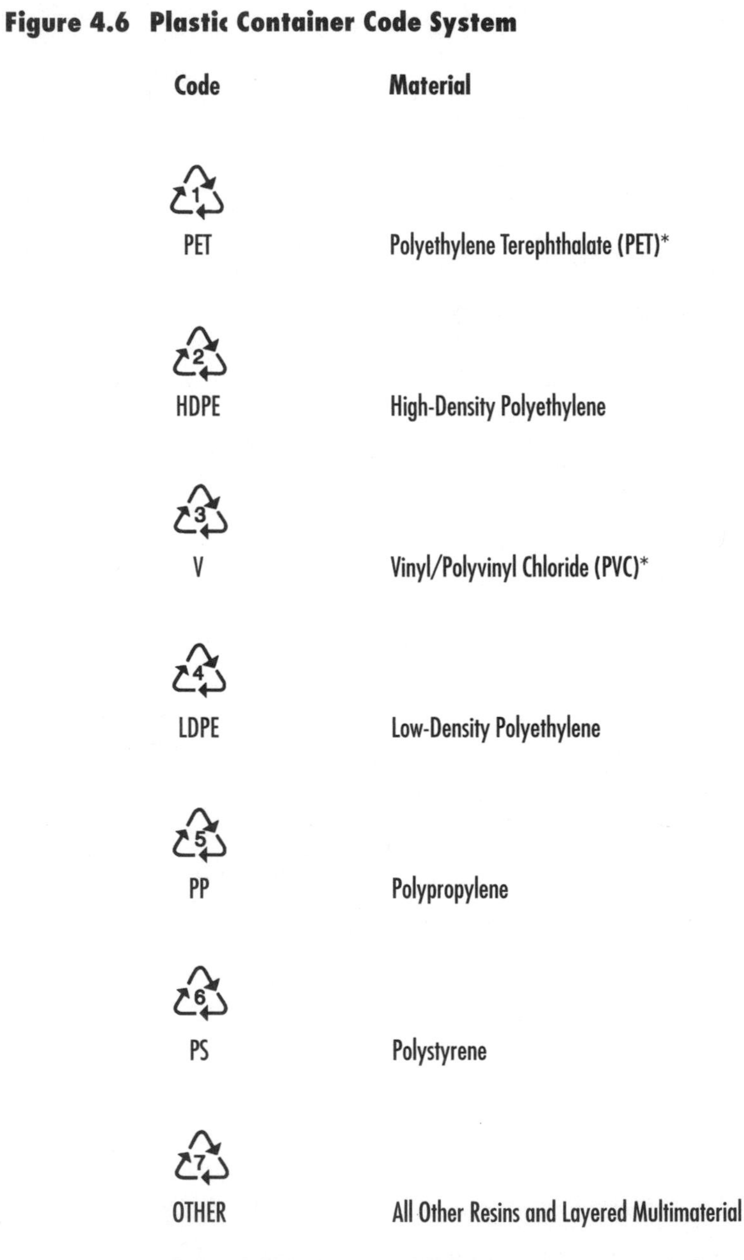

Code	Material
1 PET	Polyethylene Terephthalate (PET)*
2 HDPE	High-Density Polyethylene
3 V	Vinyl/Polyvinyl Chloride (PVC)*
4 LDPE	Low-Density Polyethylene
5 PP	Polypropylene
6 PS	Polystyrene
7 OTHER	All Other Resins and Layered Multimaterial

* Stand alone bottle code is different from standard industry identification to avoid confusion with registered trademarks.

Source: Society of the Plastics Industry, Inc.

After being collected and sorted, each type of resin is treated in a similar manner. As with glass and aluminum recycling, industry specifications for scrap plastic are set within strict limits. Very little contamination from other commodities, including other resins, is allowed. Consequently, inspection for cleanliness is the first step in any processing plant. The plastics are then compressed into bales that weigh over 1,000 pounds each. The recycling facility subsequently sends these bales on to its buyer/processor.

At this new location, the plastic is washed to remove residues from former uses. Paper labels, metal rings or caps, and other contaminants are also removed at this time. Next, the plastic is shredded, dried, and then processed further into pellets or flakes, which form the new raw materials for the next plastic product.

Plastics recycling is usually not a closed-loop process, for soft drink bottles are typically not reprocessed back into soft drink bottles (because they may have been used by a consumer to hold nonfood products, such as motor oil). However, bottling companies are working with the plastics industry to produce new soft drink bottles containing at least 25 percent recycled resin.

In general, the recycling of plastics is not nearly as straightforward as that of glass or aluminum due to the variety of resins and the light weight of the products. These two factors make plastics recycling neither cost effective nor energy efficient. For these reasons, some people in the field would like to see the entire plastics industry evolve out of existence.

Others, such as J. Winston Porter, former assistant administrator for solid waste and emergency response for the EPA, believe that plastics should be handled in three ways: (1) recycled in some cases, if it is relatively simple to do so (as with PET and HDPE); (2) landfilled in other instances because the inert properties of most plastics keep them from polluting the air or water while they sit underground indefinitely; and (3) burned in waste-to-energy plants where their high energy content can be used as a resource to produce electricity. (Plastics have about twice the heat energy of coal.)

In the early 1990s, the Eastman Chemical Company was investigating the possibility of placing molecular markers in the various plastic resins to make them identifiable by sorting machines that could "read" the marker and sort mixed plastics at a rate faster than hand sorting. When this technology was taken to its farthest ability, however, bottles could only be sorted at a rate of one every three seconds, which was not deemed fast enough for recycling needs.

However, another technique was discovered through this research and a type of spectrascope was designed that could identify the type of plastic resin that a bottle was made of in one-tenth of a second. According to Eastman, one company, Magnetic Separation Systems, Inc., in Nashville, Tennessee, has fully developed this automatic sorting method and has named it the *BottleSort*® system. Magnetic Separation Systems now markets this system throughout the world.

In spite of the advances in recycling, many environmentalists contend that there is still a basic problem with plastics production and recycling in general: The plastics industry uses five of the top six chemicals that the EPA lists as "most hazardous waste" during the initial production of virgin plastics. But when all is said and done, both industry and environmental factions agree that by recycling plastics, the initial production toxicity can be reduced because less virgin resin is needed in the long run.

Recycling the Most Common Plastic Packaging Types

PET (Polyethylene Terephthalate) ♳

PET (also known as PETE) comprises 9.1 percent of plastic packaging. This is the type of plastic most often used to make transparent soft drink and liquor bottles, as well as peanut butter jars. It is the most recycled plastic in the United States, and the only plastic that can retain carbonation.

The largest use for recycled PET is in fiber applications. About 50 percent of all polyester carpeting produced each year includes recycled content. Some companies even manufacture carpeting made of 100 percent recycled PET soft drink bottles. Other fiber uses include fiberfill for jackets, gloves, sleeping bags, pillows, and cushions. Polyester upholstery and clothing, as well as geotextiles used in roadbeds and landfills, may also contain recycled fibers. In addition, recycled PET can be found in auto parts, paint bristles, and industrial strapping.

Procter & Gamble began incorporating recycled PET into nonfood containers in 1989, and Coca-Cola and Pepsi began using it in their beverage bottles in 1991.

When recycling PET, double-check for the code number 1, because another resin type, vinyl, looks very similar but is coded with the number 3. Even a small amount of vinyl in a PET batch can ruin the entire mixture as well as the machinery.

HDPE (High-Density Polyethylene) ♴

HDPE comprises 29.4 percent of plastic packaging. This type of plastic is typically used to make the common one-gallon milk and water jugs. The HDPE in these products is transparent but thicker and less clear than PET. Detergent, bleach, motor oil, and some margarine and yogurt containers are made of HDPE in its colored and more rigid, opaque form. Some plastic grocery sacks are also made of this resin.

Processors of HDPE are generally more selective about the types of containers they buy. Clear containers are almost always accepted by plastics recycling programs, as are colored bleach and detergent bottles. But the smaller, more variable containers that hold products such as shampoo, suntan lotion, yogurt, margarine, or hair spray, as well as HDPE plastic bags, are often excluded. It is important to check with local recycling facilities for the acceptable containers in any specific area.

Most HDPE is recycled back into bottles or similar containers. Unfortunately, because the color of recycled HDPE cannot be controlled, the bottles containing some percentage of it are still covered with a layer of virgin plastic.

Other end products from recycled HDPE include many of the nesting recycling containers used in curbside recycling programs, as well as flowerpots and traffic cones.

Vinyl ♵

Vinyl comprises 4.6 percent of plastic packaging. Vinyl bottles can look quite similar to those made of PET, for they are also transparent. But vinyl bottles are a bit less rigid and are often used to package cooking oil, water, and shampoo. The more specific name for this type of plastic is polyvinyl chloride, or PVC. Water pipes, garden hoses, credit cards, and shower curtains are among the many products made of PVC.

Vinyl is clear when molded, but unlike PET, a white crease will appear where it is folded. The difference is critical, for the inclusion of even a few vinyl bottles in a batch of PET can destroy recycling equipment. When vinyl gets too hot, it produces hydrochloric acid, which eats away any chrome that might be present. When collecting clear plastic bottles for recycling, it is important to double-check for the code numbers 1 or 3.

It is not easy to find recycling programs that will accept vinyl at this time. Some recycled vinyl can be made into mats for use in truck trailers, auto interiors, and weight rooms.

LDPE (Low-Density Polyethylene) ♴

LDPE comprises 32.5 percent of plastic packaging. Products made of LDPE include shrink-wrap, dry cleaner bags, plastic sandwich bags, and some plastic grocery bags.

Technically, plastic bags can be recycled in a closed-loop system, bags to bags, and some recycling programs do accept them. More and more dry cleaners and grocery stores have collection boxes for these bags, but most LDPE is still thrown away, making up 40 percent of our plastic trash. To combat this problem, the Plastic Bag Association was founded and assists consumers looking for places to recycle bags through their toll-free number, 1-800-438-5856 (see Chapter 6 for more information).

Sticky "cling" wraps are not LDPE and are not recyclable because of the ingredient that causes them to stick.

PP (Polypropylene) ♵

PP comprises 10.3 percent of plastic packaging. Although it is technically possible to recycle PP, it is difficult to find a program that will accept it. Of the 3.5 million tons used each year in the United States, only about 1 percent is being recycled.

PP is the material used to make most plastic tops and lids for containers. Other products made of PP include maple syrup bottles, automotive battery casings, and disposable diaper linings.

When it is accepted for recycling, PP is usually mixed with other resins and made into such items as squeegies and road marker reflectors.

PS (Polystyrene) ♶

PS comprises 10.7 percent of plastic packaging. Polystyrene is one of the most controversial plastics around due to its connection with chlorofluorocarbons (CFCs), chemicals notorious for their large role in the destruction of the Earth's protective ozone layer. The U.S. foam food service packaging industry voluntarily stopped using fully halogenated CFCs in 1988.

PS also received a great deal of publicity when McDonald's Corporation decided to discontinue both using and recycling it after study and consultation with the Environmental Defense Fund in 1990.

PS is manufactured in two forms: foam and rigid. The foam may often be called by Dow Chemical's trade name of "Styrofoam™," but its correct generic name is polystyrene foam. In this form, PS is injected with gases in order to puff it up. CFCs were originally used for this purpose, but today, either pentane or carbon dioxide is used by most manufacturers. HCFC-22, another type of chlorofluorocarbon that contributed to atmospheric problems, was phased out of use in most foam applications on 1 January 1994.

Polystyrene foam is commonly found in packing "peanuts," coffee cups, carryout food boxes, and insulating materials. The rigid form is found in some yogurt and sour cream tubs, as well as salad bar carryout containers. To determine if a product is polystyrene, look for the code number 6 on the bottom of the item.

6

The largest reuse program for polystyrene foam "peanuts" has been undertaken by more than 3,500 Mail Boxes, Etc. and other shipping stores. The Polystyrene Packaging Council is also working to expand PS recycling opportunities (see Chapter 6 for more information). Prior to 1988, there was virtually no postconsumer recovery of PS, but in 1994 more than 34 million pounds of PS were recycled.

Recycled PS can be made into videocassette tape boxes, combs, reusable cafeteria trays, license plate frames, trash cans, notepad and desk calendar holders, flying discs, rulers, and yo-yos. It is not made back into cups or packaging and therefore is not recycled in a closed-loop system. This means that even if all the PS used were being recycled, production of virgin PS would not be affected in the least and the same amount of raw materials and toxic wastes would still be produced.

Other Plastics 7

Miscellaneous plastics comprise 3.4 percent of plastic packaging. Some of these plastics contain a mixture of the resins discussed earlier; others are made of entirely different resins.

One process by which mixed resins were first recycled together was developed around 1990 by the Center for Plastics Recycling Research at Rutgers University and is known as ET-1. The end product of the ET-1 mixed plastics recycling process was the first type of plastic lumber used in park benches and picnic tables.

More recently, one U.S.-based plastics recycling company, Hettinga Technologies, Inc., has taken the ET-1 idea a few steps further, mixing all types of plastic resins collected from consumers, clean or contaminated, and producing products meeting uniform specification standards. By its unique and patented process called Controlled Density™ (CD) Molding, this company is taking the materials that others would still discard as garbage and creating highly usable products such as marine plywood, industrial furniture, concrete forms, roofing materials, and dog kennels. The plastic construction materials produced by Hettinga can be used structurally like wood but are immune to damage from water, chemicals, bacteria, insects, and rodents. It does not conduct electricity, does not rust, and will not splinter or split. The advantage it has over earlier forms of plastic lumber is its foam core, making it lighter (less expensive to transport) and easier to work with.

Paper

Paper Facts

- Every day, U.S. paper manufacturers recycle enough paper to fill a 15-mile-long train of boxcars.
- In 1995, recovered paper provided one-third of all the raw material fiber used in U.S. mills (up from 25 percent in 1988).
- Each year, 2 billion books, 350 million magazines, and 24 billion newspapers are published in the United States.
- In 1995, 43.3 million tons of paper and paperboard were recovered in the United States (21 million more tons than in 1986).
- More than one-third of all the paper and paperboard recovered in the world is recovered in the United States.
- In 1995, about 329 pounds of paper were recovered for each person in the United States.
- Every day, Americans now recover over 200 million pounds of paper.
- About 10 million fewer tons of paper were landfilled in 1994 than in 1987, despite the significant increase in paper consumption during that time.
- As of 1994, more paper was being recovered than was sent to landfills.
- Approximately 80 percent of all the paper recovered in the United States is recycled into new paper and paperboard

products; the remainder is either exported to foreign recyclers or reused in this country to make products such as animal bedding, hydromulch, and compost.

- Between 1986 and 1994, recovery of all paper grades has increased, and total recovery has grown nearly four times faster than the growth of paper and paperboard consumption.
- Americans now recover more than 40 percent of all office papers.
- Nearly 70 percent of all containers and packaging recovered in the United States are made of paper or paperboard.
- According to the Environmental Protection Agency, 44 percent of all paper/paperboard packaging is currently being recovered, compared to about 20 percent of all other packaging materials combined.
- Nearly 6 out of every 10 corrugated boxes are now being recovered from the waste stream.
- In 1995, 70 percent of all corrugated cardboard was recovered (an all-time record).
- Since 1990, the recovery of high-grade office paper has increased more than 50 percent.
- One year's worth of the *New York Times* weighs 520 pounds.
- In 1994, more old newspapers were recovered in the United States than newsprint producers manufactured domestically.
- While 33 percent of all the newsprint used in the United States was recovered in 1986, by 1994 this recovery rate had risen to 59 percent.
- Since 1990, domestic papermakers have increased their consumption of old newsprint by one-third.
- Every year, Americans use more than 75 million tons of paper and paperboard products (about 600 pounds per person).
- Every day, U.S. businesses generate enough paper to encircle the Earth 20 times.
- Approximately 70 percent of office trash is recyclable paper.
- The paper industry is the single largest user of fuel oil in the United States.
- Manufacturing recycled paper uses up to 64 percent less energy than manufacturing virgin paper.
- Manufacturing recycled paper reduces water pollution by 35 percent.
- Manufacturing recycled paper reduces air pollution by 74 percent.
- Manufacturing recycled paper uses 58 percent less water.
- The collection and recycling of paper provides five times as many jobs as the harvesting of virgin timber.

- Every ton of paper that gets recycled saves approximately 17 trees.
- Every ton of paper that gets recycled keeps 60 pounds of pollutants out of the air.
- Every ton of paper that gets recycled saves 4,100 kilowatt hours (enough energy to heat the average home for six months).
- Manufacturing one ton of recycled paper uses 7,000 gallons less water than manufacturing one ton of paper from virgin pulp.
- The average office worker uses about 180 pounds of high-grade paper (white bond and computer paper) every year.
- In 1995, paper comprised about 33 percent of the waste disposed of in U.S. landfills, even after recycling.
- Americans make about 750,000 photocopies every minute of every day.
- The average 100-person company uses about 378,000 sheets of copier paper per year, an amount that would make a stack about seven stories high.
- By photocopying on both sides of a piece of paper, paper usage can be cut by 50 percent.
- One pulp tree can produce about 12.5 reams of paper.
- Thermographic (shiny) FAX paper is not recyclable.
- If everyone who owns a FAX machine used a half-page cover sheet instead of a full page, the United States could save about 2 million miles of unrecyclable FAX paper in one year.
- If one out of every ten FAX users switched to a plain-paper FAX machine, we could save 500,000 miles of paper each year.
- In a recycling program, white paper is worth more than colored paper.
- Each ton of paper that gets recycled saves 3.3 cubic yards of landfill space.

Making and Recycling Paper

The technology used to manufacture recycled paper is one of the oldest in the recycling industry. Indeed, paper produced from recycled textile fiber was the only kind of paper available in the United States from 1690 to the mid–nineteenth century.

Then the demand for paper rose so sharply that there was not enough recycled fiber to supply the industry. Paper producers therefore created a technology to use wood pulp from trees to supplement their fiber needs. Eventually, wood fiber became the material of choice for papermaking because of its strength, and it supplanted the use of recycled fibers.

All paper is made in essentially the same way, whether it is recycled or not. It is made of pulp, a mixture of pulverized plant fibers and water. In the United States, most virgin pulp (that is, pulp that does not contain recycled fibers) is made from the wood of trees harvested specifically for this purpose. Once the trees have been chopped, shredded, and mixed with water to make pulp, the papermaking process begins.

There are two basic types of commercial papermaking facilities: the paper mill and the board mill. In a paper mill, the pulp is placed in a head box that controls the amount of fiber that is distributed onto a wire screen. This screen, which can be hundreds of feet long, skims through the head box, catching and holding the fibers from the pulp while allowing the water to drain away. The wire screen travels a complicated path, taking the relatively uniform layer of fibers through a series of rollers that squeeze out the excess moisture. When most of the moisture has been removed, the fibers have become pressed into a continuous sheet of paper. This sheet is passed through more hot rollers that remove the last traces of moisture and give it a smooth finish. Finally, the paper is collected onto huge rolls to be sent to buyers.

In a board mill, the process is similar, but instead of using a head box to distribute pulp fibers, a number of cylinders below the wire screen deliver the fibers to the wire. A board mill can add to or decrease the number of cylinders, depending on the thickness and color desired in the finished product. For instance, to make the type of paperboard found in a cereal box, several cylinders of unbleached, brown-gray pulp make up most of the layers, and the final cylinder holds bleached, white fibers that form the top layer, which will later contain printing.

The major difference between virgin and recycled papers is the pulp. Virgin paper is made from virgin pulp, processed directly from trees. Recycled paper is made from recycled paper pulp, also called *slurry,* which is made by mixing the collected paper with water and mechanically beating the mixture in a hydropulper (a machine similar to a giant blender) until the fibers are once again separated.

Another difference in the manufacturing of some recycled papers is the de-inking process. For example, newspaper that has been collected from consumers in recycling programs is covered, for the most part, with printing inks. And before the fibers from these papers can be reprocessed into new paper, the ink must be removed.

Until the relatively recent resurgence of interest in using recycled paper products, the few de-inking plants that were operating in the United States were just sufficient to handle the available supplies of recycled paper. When recycling became popular again in the late 1980s and the number of U.S. collection programs began growing almost exponentially, there was a period when these facilities could not handle the supply.

Consequently, in the late 1980s and early 1990s, there was a lag between the popularity and success of recycling and the building of new de-inking plants. The sudden glut of newspaper was the reason why the buy-back prices for this product dropped so precipitously at that time; there was simply much more newspaper being collected than could be de-inked and reprocessed. Additionally, the demand for recycled newsprint was not high enough to encourage investors to risk their capital on the expensive venture of de-inking plants.

However, with the ensuing advent of legislation in an increasing number of states that required certain percentages of recycled fibers to be used by newspapers, numerous de-inking plants were planned for construction and began operating in the mid-1990s. In fact, since 1990 domestic papermakers have increased their consumption of old newsprint by 33 percent, and by 1995 there were 21 paper mills operating in North America that bought recovered newsprint for recycling. According to the American Forest & Paper Association, many more mills, specifically designed to handle recovered paper for recycling, are currently being planned or are actually under construction.

By manufacturing paper from recycled fibers, the need for virgin pulp can be greatly reduced. However, this pulp is not likely to be totally eliminated from papermaking because paper fibers can only be recycled a finite number of times. Each time they are recycled, they are cut or broken down a bit more, and shorter fibers create papers with less strength and lower quality. For this reason, most recycled paper still contains either some virgin pulp fibers or fibers derived from wastepaper with longer fibers. (Recycled paper now comes labeled as to how much postconsumer recycled fiber it contains.)

Addressing the use of virgin wood pulp is another aspect of the recycled paper industry. At the same time that recycling was being promoted for the sake of conserving resources, environmentalists were decrying the deforestation of our country for the sake of paper products. In the ensuing years, many new options have been, and continue to be, explored for the sake of finding paper fibers that can be used without environmentally destructive consequences.

Supplementation or replacement of wood fibers are being found in the forms of other, more sustainable agricultural products, such as hemp or kenaf. But a new alternative, banana paper, is becoming more popular. Banana paper is produced from a mixture of mostly recycled paper fiber with the addition of fiber from banana stalks. These stalks are the agricultural by-product of banana plantations and would normally be discarded into landfills or dumped by rivers in Central and South America. This successful new technology, first created by students at a Costa Rican university and a paper manufacturer who cooperatively formed the Costa Rican Natural Paper Company, is combining both source reduction and recycling techniques to produce an innovative, environmentally sound product. The company estimates that using one ton of banana fiber saves an estimated 17 trees. Banana paper products are currently available at Kinko's™ and by mail order through several eco-retailers in the United States.

What Is Recycled Paper?

Distinctions between grades of recycled paper are based on fiber content. Some brands are labeled "100 percent recycled paper" but contain only preconsumer waste. This is waste generated from the mill broke, or leftover scraps of a papermaking plant; the material has never been used by a consumer.

Other "100 percent recycled paper" brands contain a certain percentage of preconsumer waste fibers as well as postconsumer waste fibers, generated from paper that has been used by consumers and collected in a recycling program. Interestingly, board mills have always made paperboard from 100 percent recycled fibers, both pre- and postconsumer, and they do not de-ink. Thus, printing the "recycled" symbol on paperboard products is a marketing tool geared toward the modern "green" consumer.

At this time, the term *recycled paper* has no specific, legally binding definition. The National Recycling Coalition and the Environmental Protection Agency are continuing to work on the issue, but it is a highly controversial one in the paper industry. Although there are specific EPA guidelines on recycled paper, they apply only to purchases made with federal monies.

The EPA guidelines concerning printing and writing papers recommend that a paper contain at least 50 percent wastepaper in order to be considered recycled. The definition of wastepaper, however, is quite broad, including both pre- and postconsumer wastes. In some instances, the inclusion of fiber made from sawdust qualifies as recycled content.

For anyone other than federally funded agencies, paper that is labeled as recycled can contain pre- or postconsumer waste in any imaginable quantities. For this reason, many states are creating their own guidelines for recycled paper procurement, based on specified amounts of each type of fiber. (These state guidelines are outlined in Chapter 5.) This is a confusing and debilitating problem for the paper industry because almost any paper can be called "recycled" according to one definition or another. As a result, manufacturing recycled paper to encompass every purchaser's specifications has become a difficult business.

Collecting Paper for Recycling

For all practical purposes, any paper goods that can be made from virgin fibers can also be made from recycled fibers. Among the products manufactured from recycled paper and paperboard collected from consumers are corrugated boxes, newspaper, printing and writing papers, greeting cards, tissues, paper towels, cereal boxes, beer cartons, tissue boxes, insulation, acoustic ceiling tiles, gypsum wallboard, flooring, egg cartons, fruit trays, cement bags, shoe boxes, and grocery sacks. However, collecting the types of paper necessary to make these products can be confusing.

There are many different kinds and grades of scrap paper (at least 51 regular grades and 33 specialty grades, according to the Institute of Scrap Recycling Industries and the American Forest & Paper Association), but they can all be broadly categorized into two varieties depending on the type of fibers they contain: groundwood and free sheet.

The paper grades most commonly recycled by consumers include Sorted White Ledger #40, Computer Printout (CPO) #42, Groundwood CPO #25, Newspaper #8, and Corrugated #11. Together they form a very large proportion of the paper recycled by the general public in curbside, buy-back, and office paper recycling programs.

As with other recyclable products, paper collected for recycling must be free of contaminants. It must also be free of other grades of paper whose fibers will contaminate the recycled paper products to be made. These contaminating grades of paper are called out-throws. Descriptions of the major paper grades, together with their commonly occurring contaminants and out-throws, follow. Local recycling centers or scrap paper buyers should be consulted for other specifications that may exist.

Groundwood

Groundwood is the cheapest and most plentiful of the paper types. It includes paper made from fibers that have been manufactured solely by mechanical action, with no chemicals involved. Because no chemicals are used, groundwood contains a naturally occurring substance, normally found in trees, called lignin.

Lignin is the ingredient responsible for making paper turn yellow and brittle with age. It is also the easiest substance for paper recyclers to test for when they are trying to determine a particular paper's grade. By using the acid phloroglucinol, which reacts to the presence of lignin by turning purple, one can determine whether a paper contains groundwood fibers. This test is often referred to as the "acid test."

Because of its lignin content, groundwood paper is generally used for printed materials that are not meant to be stored in archives but are intended to be used briefly and then discarded.

Some groundwood paper is coated, and some is sold uncoated. Uncoated groundwood is used in the majority of newspapers in the United States. Coated groundwood, which has a shiny surface layer of clay (usually Georgia Kaolin clay) to improve its printing surface, is most often found in magazines and catalogs.

Free Sheet

Free sheet paper is made of fibers that have been chemically treated to remove all lignin. This type of paper can be further categorized as unbleached and bleached, according to how the fibers are treated.

Unbleached free sheet refers to paper that is free of lignin but has not been whitened or lightened from its natural brown color. This is the type of paper most often found in corrugated cardboard boxes and grocery sacks.

In bleached free sheet, the fibers have been chemically lightened and whitened either with chlorine or with hydrogen peroxide. It is noteworthy that more and more paper manufacturers are switching away from the chlorine-bleaching method because of the environmental and health hazards associated with the chemical dioxin, a toxic by-product of the chlorine-bleaching process.

European countries no longer buy paper products that have been bleached with chlorine.

Most of the paper that is used for printing materials other than newspapers and popular magazines is a variety of bleached free sheet, which can be further broken down into two more categories—uncoated and coated—according to whether the paper is coated with a substance meant to improve its printing surface. Coatings make a paper smoother so that it will pick up inks more neatly. Coated, bleached free sheet is generally smooth, shiny, and more rigid than magazine paper; it is often used for printing high-quality brochures and posters.

Sorted White Ledger #40

This is a high-grade paper that is free sheet, bleached, and uncoated. It is the type of paper usually found in offices as bond, xerographic, and 9½ x 11-inch white, fanfold computer paper. It is acceptable for recycling with either impact printing (from a typewriter or ink printer) or laser printing on it, as well as pencil, xerographic carbon, and other writing inks. It may have staples or paper clips attached, but it cannot have crayon on it.

Common contaminants include glass, plastic, metals, foils, carbon paper, non-water-soluble adhesives, tape, self-sticking notes, food, candy wrappers, coffee grounds, and assorted other "trash."

Out-throws include colored paper, coated and chemically treated paper (including ream wrappers and "no carbon required," or NCR, paper), CPO, groundwood CPO, manila file folders, newspaper, and cardboard.

Computer Printout (CPO) #42

This is another high-grade paper that is free sheet, bleached, and uncoated. It is fanfold computer printing paper and generally has pastel-colored horizontal bars (most often green). This paper is typically used with larger-than-letter-size computer printers. It is acceptable for recycling when printed or written on in ink or pencil, but it cannot be printed on with a laser printer or with crayon.

Contaminants are the same as those for sorted white ledger, but paper clips and staples are also considered contaminants.

Out-throws include sorted white ledger paper, colored paper, coated and chemically treated paper, groundwood CPO, manila file folders, newspaper, and cardboard.

Groundwood CPO #25

This is a low-grade paper that basically looks like CPO but contains groundwood. In better-quality brands, it is virtually indistinguishable from CPO #42. For this reason, the acid test is used to determine its identity. It is acceptable if impact or laser printed, but it must not be colored with crayon.

Contaminants are the same as for CPO #42.

Out-throws include sorted white ledger paper, colored paper, coated and chemically treated paper, CPO, manila file folders, newspaper, and cardboard.

Newspaper (ONP) #8

This is another low-grade, groundwood, uncoated paper that has been printed on and is used by most U.S. newspapers. For recycling, the glossy, colored inserts (Sunday magazines, coupons, and advertising pages) that are delivered inside the paper may be included. But newspaper that has become yellowed by age or sunlight should be excluded because the chemical reaction in the lignin within the paper fibers, which is responsible for the color change, degrades the paper.

Contaminants are the same as for CPO #42.

Out-throws include paper grocery sacks, plastic delivery bags, sorted white ledger paper, colored paper other than inserts, coated paper other than inserts, CPO, groundwood CPO, manila file folders, cardboard, telephone directories, mail-order catalogs, magazines, and junk mail. Some recycling programs and scrap paper buyers will accept a limited amount of groundwood CPO with newspaper, but local recycling programs should be consulted for more information.

Old Corrugated Containers (OCC) #11

This is a low-grade, free sheet, unbleached, and uncoated paper. Commonly called "corrugated cardboard," it is often used to make boxes. It is distinguished from other cardboards by its three-layered composition: two outer liner layers, with the wavy medium layer sandwiched between them.

Contaminants include plastic, glass, metals, foils, carbon paper, food, coffee grounds, and assorted other trash.

Out-throws include paper grocery sacks, sorted white ledger paper, colored paper, CPO, groundwood CPO, manila file folders, noncorrugated cardboard (such as beer cartons, shoe boxes, and cereal boxes), telephone directories, mail-order catalogs, magazines, and junk mail. Some recycling programs and scrap paper buyers will accept brown paper grocery bags with OCC, but local recycling programs should be consulted for more information.

Other Papers

Used Brown Kraft #15, another type of paper most consumers are likely to be familiar with, is collected for recycling in some areas. Like OCC, this paper is low grade, free sheet, unbleached, and uncoated. It is primarily found in brown paper grocery bags, which are often collected for reprocessing after being used by the consumer. As with other paper grades, very little contamination is permissible.

Another common grade of paper that is more widely recyclable now, but was not in the past, is Coated Groundwood, which is most often found in magazines. Whereas few programs used to accept magazines for recycling (because when this type of paper is repulped, it produces more clay than paper fiber), new technologies have been developed that use clay in the de-inking process. This new process is known as flotation de-inking.

As facilities using the flotation de-inking method became more widespread, recycling programs have found new markets for magazines and similar types of paper. Somewhat as a result of this, a new grade of recovered paper was created and called Mixed Papers (this grade was also born of the need to find a market for all the varieties of paper that consumers were using but that were not acceptable for recycling). This new grade includes a mixture of almost any paper not covered by the other specifications, such as colored ledger paper, magazines, catalogs, and telephone directories.

A common contaminant in many paper grades is non-water-soluble adhesive. This includes hot glue, self-stick labels, pressure-sensitive labels and closures, and all self-stick tapes. These contaminants are called "stickies" by the scrap paper industry.

Stickies are one of the most costly contaminants in recyclable paper. Their presence is destructive in two ways: (1) They can clog the pulp preparation cleaning equipment because they normally do not dissolve in water, and (2) if the water is hot enough, they may dissolve, only to solidify later in the papermaking process, causing the paper to stick to the wire or to itself as it is being rolled, causing rips and tears.

Due to the fine distinctions between paper grades, it is imperative that anyone establishing a recycling program consult local markets for contamination specifications. Additionally, the people who are actually collecting the paper must be taught to distinguish between contaminating grades and acceptable ones. Without this knowledge, many paper recycling programs fail because it is too expensive for haulers and recycling processors to separate out the trash.

Food and Yard Wastes: Composting

Composting Facts

- In every gram of compost, there are 1 billion organisms.
- There are 20 million acres of grass in the United States—the equivalent of a lawn as big as the states of Vermont, New Hampshire, Massachussetts, and Rhode Island, combined.
- In 1994, almost 15 percent of our solid waste was comprised of mowed grass, dead leaves, and branches.
- Americans generated about 31 million tons of yard waste in 1994.
- Between 50 and 80 percent of the nutrients absorbed by a tree are contained in its fallen leaves.
- By recycling one family's yard waste for one year, between 300 to 400 pounds of compost, or *humus*, can be produced.
- When yard waste is buried in a landfill, where it is compacted and receives little oxygen, its decomposition produces methane, a greenhouse gas.
- Americans throw away the equivalent of over 21 million shopping bags full of food every year.
- Americans throw out about 870,000 pounds of food every day.
- Each day, over a third of the 238 pounds of waste produce by an average McDonald's restaurant is food.
- In 1994, Americans generated over 14 million tons of food waste.
- Food comprised about 7 percent of the waste generated in the United States in 1994.
- In 1989, the United States had over 1,000 municipal composting programs; by 1995, there were over 3,000.
- Americans throw away about 10 percent of the food they buy at supermarkets.

Composting Food and Yard Wastes

Food and yard wastes are major components of U.S. landfills, composing approximately 21.3 percent of all solid waste. As of 1995, there were over 3,200 yard waste composting programs in the United States, and 23 states had banned these wastes from landfill disposal. Composting is the aerobic (oxygen-dependent) process by which plant and other organic wastes decompose under controlled conditions. It is among the most natural recycling processes on Earth, since this is the same basic process by which dead leaves, grasses, and animals decompose in the wild. When humans compost, however, they generally mix wastes in such a manner as to speed up the process, producing humus, a rich, crumbly type of soil.

People have been composting for thousands of years. The process is mentioned in the Bible more than once, and ancient agriculturists used compost to enhance their soils just as modern farmers are relearning (and improving upon) the art and science of composting. Over 2,000 years ago, Marcus Cato, a Roman farmer/scientist, wrote about the necessity of composting food and animal wastes to make the soil more fertile.

Currently, large cities are using municipal compost projects and encouraging backyard composting and vermicomposting (the use of worms to compost) to cut down on the amount of organic wastes being disposed of in landfills.

The Science of Composting

Composting is the scientific way of taking the naturally occurring process of decay and enhancing it, or speeding it up. In order to produce high-quality humus, it's important to understand the way various biological factors interact. Simply throwing a bunch of food and yard wastes in a pile is "cold" composting, and while it will eventually decay, it is likely to be a long and smelly process. "Hot" composting, the version utilized by most people who have compost programs, is quicker but takes more work.

In order for organic material to decompose, two processes must occur: one physical, the other chemical. Physically, the appetites and activities of invertebrates such as earthworms, mites, sowbugs, millipedes, and beetles break down the large pieces of food and yard waste into smaller pieces. Chemically, soil microorganisms such as fungi, some protozoans, bacteria, and actinomycetes break down larger, more complex organic compounds into simpler compounds, as well as carbon dioxide and water. Another of the byproducts of this chemical breakdown is heat (hence the term "hot" composting"), which is why many composters use thermometers to monitor the health of their compost piles.

In order for all these processes to occur, oxygen must also be present (which is why, in modern, compacted landfills, very little decomposition takes place). Other necessities for optimum results include adequate sources of carbon and nitrogen. Thirty parts carbon to one part nitrogen is considered the "perfect" ratio. Mixing organic wastes in the right amounts to achieve this ratio is the "art" of composting.

The Art of Composting

Yard wastes and food scraps can be classified in two groups, "green" and "brown." "Green" materials include fresh grass clippings, kitchen scraps, and garden plants, while "brown" materials include dried leaves and branches. The green substances contain large amounts of nitrogen, while the brown substances are relatively low in nitrogen but quite high in their carbon content.

Finding the right "recipe" for compost varies according to climatic condition, such as temperature and humidity, as well as the mixture of green and brown materials. Books, periodicals, and Internet World Wide Web sites all contain information on how to get the mixture "just right," according to where one lives and the nature of one's compost materials (see Chapters 7 and 8).

Adequate moisture, 40 to 60 percent by weight, needs to be maintained, and the essential ingredient, oxygen, can be added by simply mixing the pile regularly with a pitchfork or other such implement. (Composting can take place without oxygen, or "anaerobically," but this method produces numerous malodorous compounds such as methane and hydrogen sulfide gases, so it is not the most popular way.)

Once the decaying process has finished, the end result is the odorless, porous and crumbly, nondegradable organic mixture known as humus. As a soil conditioner, humus/compost improves the ability of soil to retain water and nutrients, improves its buffering capacity, and aids drainage in clay soils while increasing water retention in sandy soils.

Vermicomposting

Using worms to compost organic wastes is known as vermicomposting. This method of composting is becoming increasingly popular in urban areas, as one can take food wastes and compost them in a relatively small area, a worm box or bin, indoors or out on a patio if the weather is warm enough.

Worm bins and worms, usually the species *Eisenia foetida* (red worms), are available through mail-order, and the only other materials needed are food scraps and bedding, such as shredded newspapers. In vermicomposting, the most important factor for success is the ratio of food scraps to worms to surface area in the bin.

The final product of this type of composting is worm "castings," another nutrient-rich fertilizer for gardens or potted plants. Again, books, periodicals, and Web sites on the subject cover the details and offer troubleshooting information on vermicomposting (see Chapters 7 and 8).

Metal

Scrap Metal Facts

- Steel is the world's most recycled material.
- In the United States alone, over 70 million tons of steel were recycled in 1995.
- More than 12 million cars were recycled in 1995—if you lined that many cars up, it would cause a traffic jam stretching from New York City to Los Angeles 15 times.
- In 1995, 13 million tons of steel were recovered from automobiles (800,000 tons of this steel went back to the auto industry to produce new cars).
- The amount of steel recovered and recycled from autos in 1995 (13 million tons) is enough to build about 151 Golden Gate bridges.
- Each year, North Americans abandon 3 million cars.
- Aluminum was worth more than gold when it was first discovered.
- North Americans throw out enough steel and iron to continually supply all the nation's automakers.
- Operating at full capacity, the U.S. scrap industry has the capacity to process 140 million tons of iron and steel annually.

- Steel made from scrap is chemically and metallurgically equivalent to steel made from virgin ore.
- More than half of all steel manufactured in the United States today is made of recycled material.
- Recycling ferrous metals saves 74 percent of the energy used to make them from iron ore and coal.
- Making steel from scrap saves 90 percent of the virgin, nonrenewable materials used in making steel from ore.
- Recycling steel and iron reduces air pollution by 86 percent.
- Recycling steel and iron uses 40 percent less water.
- Recycling steel and iron produces 76 percent less water pollution.
- Every ton of steel that gets recycled saves 2,500 pounds of iron ore, 1,000 pounds of coal, and 40 pounds of limestone.
- Recycling steel and iron reduces mining wastes by 96 percent.
- It takes four times as much energy to make steel from virgin iron ore as it does to make it from scrap.
- Each year, steel recycling saves enough energy to provide electricity to about one-fifth of the households in the United States (or about 18 million homes) for one year.
- At current landfill tipping fee rates, recycled steel saves the United States over $2.5 billion per year in avoided solid waste disposal costs.
- The amount of steel recycled each year equals about one-third of the amount of all the municipal solid waste that is landfilled.
- Products manufactured from recycled steel and iron, virgin ore, or a combination of the two will perform identically.
- The U.S. scrap industry reprocesses enough copper each year to make 360 trillion pennies.
- Each year, the U.S. scrap industry handles enough stainless steel to make 11 billion spoons.
- The U.S. scrap industry annually recycles enough aluminum to make siding for 8 million homes.
- In 1995, 41.8 million appliances were recycled (9.5 appliances for every person who visited the Grand Canyon that year).

Making and Recycling Metals

Scrap metals can be divided into two basic categories: ferrous and nonferrous. Ferrous metals contain iron ore, and magnets are attracted to them. Iron and steel are ferrous metals.

Nonferrous metals contain little or no iron, and thus have no magnetic properties. Copper, aluminum, silver, and gold are nonferrous metals.

When metals are manufactured from virgin materials, the process is, in broad outline, similar to the process for manufacturing aluminum. Basically, the ore is mined, refined, melted in a furnace, and cast into ingots to be bought by processors to make new products. When scrap metals are collected for recycling, they are generally crushed, shredded, melted and re-refined, and poured into ingots.

There are three types of scrap metal that can be recycled: (1) obsolete or old scrap, composed of metals that have been used by a consumer, such as old cars and ships, appliances, and electrical wire; (2) industrial scrap, also known as new or prompt scrap, which is scrap produced by the manufacturing process, such as metal filings and leftover stampings from making appliance parts; and (3) home scrap, which consists of the leftover metals that steel mills and foundries produce while manufacturing new products. Home scrap is generally returned to the furnace on-site to be remelted, without ever leaving the plant.

Because of the ability of metals to be remelted, re-refined, and reprocessed indefinitely without the loss of their required characteristics, most scrap metal is recycled back into products similar to the ones originally made from virgin materials. Metals recycling is one of the most efficient ways to conserve natural, nonrenewable resources.

Ferrous Metals

Ferrous metals include iron, which is produced from the refining of iron ore, and steel, which is produced when carbon, and sometimes other elements, are added to iron. Different grades of steel result from varying the amount of carbon. Steel is the most widely used metal alloy in the world and the most widely recycled commodity.

Most ferrous scrap is bought by dealer/processors from the industry and general public and is torched, cut, baled, or shredded for further handling. Brokers then assist the dealer/processors in finding markets for their materials. They also assist scrap consumers by locating sources for the metals they require. Finally, the consumers, or mills and foundries, purchase the processed scrap and remelt it to produce their new products.

Contemporary technology produces steel in two ways, using two different types of furnaces. Both of these furnace types require old steel to make new products. The first is called the Basic Oxygen Furnace (BOF) process, and uses 25 to 30 percent old steel to make new. This process produces products such as automotive fenders, car bodies, refrigerator encasements, 5-gallon pails, 55-gallon drums, and packaging like soup cans.

The second type is called the Electric Arc Furnace (EAF) process. This method uses virtually 100 percent old steel to make new, and produces products such as structural beams, railroad ties, bridge spans, steel plates, and reinforcement bars (whose major required characteristic is strength).

It is important to note that the steel industry, unlike other recycling industries, does not rely on "recycled content" percentages to motivate scrap recovery or purchase of its products. This is because economics drive the recycling process, for recycling is simply a basic part of the business and has been for over 150 years. The environment has little to do with recycling in this area; recycled content, for steel, is an inherent part of the steelmaking process. The fact is, after its useful product life, steel is recycled back into more steel, regardless of its BOF or EAF origin.

Nonferrous Metals

Aluminum, as discussed in detail above, is produced from the refining of bauxite ore. Aluminum has been produced in commercial quantities for less than 100 years, and yet it is consumed worldwide in quantities second only to steel. It has a high strength-to-weight ratio, weighing about one-third as much as steel or copper, and as such is suited to aircraft, ships, and automobiles. Primary sources of obsolete aluminum scrap include beverage cans, cars, aircraft, appliances, furniture, and electric utilities.

Copper is produced from the refining of copper ore. It was first used around 8000 B.C. as a substitute for stone, although copper jewelry has been found dating back to 8700 B.C. For almost 5,000 years it was the only metal known to man. Currently, it is widely used for its chemical stability, electrical and thermal conductivity, and its malleability. Many common alloys are made from copper. Brass is a copper alloy containing zinc as its principal alloying element, and bronze is copper alloyed with tin and zinc. Major sources of copper scrap include car radiators, pipes and other plumbing fixtures, telephone and electrical wire and cables, electric motors, and ammunition shell cases.

Copper alloy scrap provides about half of the copper consumed in the United States each year. Most of this scrap (67.4 percent in

1995) is remelted directly by brass mills, wire rod producers, foundries, and ingot producers.

Lead is produced mostly from the refining of galena sulfide, or lead glance. It is widely used for its low electrical conductivity, its ability to absorb short wave electromagnetic radiation, and its general chemical resistance. This last quality is the reason why the largest use of lead is in the manufacture of storage batteries.

Lead is the most recycled metal, measured as a percentage of the virgin product that is reprocessed. However, because of its toxicity, not all recycling programs are equipped to handle lead safely. The major source of obsolete lead is scrapped batteries.

There are many other grades and varieties of nonferrous metals. Aluminum, lead, and copper, along with the copper alloys of brass and bronze, have many subgrades.

A nonferrous scrap metals dealer must have a thorough knowledge of metals in order to identify properly what a recycler is selling as scrap and what a consumer requires for manufacturing. After identifying and sorting scrap components, the dealer/processor prepares the scrap and packs it for shipping according to the requirements of the consumer who will be using it. It may be baled, packed in boxes or drums, or even packed loose in trucks or railroad cars.

For the average recycler, metals recycling can be simple or complicated, according to local markets. There are many grades of recyclable metals and, as with paper recycling, different markets have different specifications. Local recycling programs and scrap metals dealers should be consulted for appropriate requirements in any given locale.

Miscellaneous Recyclable Products

As public awareness about environmental issues increases and industries respond with advancing technologies, more and more products are becoming recyclable. What follows are brief discussions of some materials that are being accepted for recycling on a more limited basis across the United States.

Tires

About 253 million tires are thrown out every year in the United States. They are difficult to dispose of because they collect the methane produced in landfills and "float" up to the surface. Due to these difficulties, many landfills have refused to accept whole tires for many years, and they have ended up being stockpiled in veritable mountains, creating extreme fire hazards to surrounding communities. It is estimated that there are approximately 700 to 850 million scrap tires in stockpiles around this country.

One result of these problems has been legislation passed in many states that bans tires from landfills and requires that they be recycled and reused. Subsequently, new solutions have been developed and tire recyling/reuse has come a long way over the past decade or so. In 1985, only 10 percent of the scrap tires in the United States were being reused. By 1996, that figure had risen to 95 percent.

In 1989, about 2.5 million tires were recycled into a paving surface of combined rubber and asphalt. Currently, however, the most common way that tires are recycled in the United States is through retreading. Nearly 100 percent of the world's airlines, as well as off-road, heavy duty vehicles, use retreaded tires.

Tires not suitable for their original use or for retreading are called scrap tires. There are three major markets for these tires: tire-derived fuel (TDF), products made from pieces of rubber, and civil engineering applications.

An increasingly popular use for scrap tires is as fuel for power plants, cement kilns, and pulp and paper mills. In order to be burned, the tires are first shredded or chopped into chunks to produce TDF. The high petroleum content of TDF makes it a relatively cheap source of energy, and TDF fuel currently consumes 101 million scrap tires annually.

Other tire recycling programs shred and grind the tires to recover the two major components, rubber and steel wires and threads. The rubber can be remanufactured into products such as trailer flooring, carpet padding, and boat bumpers.

A few enterprising entrepreneurs are making fashion accessories and shoes out of used tires, while others invent new uses for them regularly. Civil engineering uses include making dams, retaining walls, and structures from scrap tires. For example, near Tucson, Arizona, at the 50,000-acre King's Anvil Ranch, scrap tires are being used to build dams to stop the flow of sediment during the rainy season.

Used Oil

As of 1996, experts estimated that Americans illegally dump an amount of waste oil equivalent to at least 35 *Exxon Valdez* oil spills each year (426 million gallons). But motor oil is a lubricant that can be re-refined over and over again, and if all the motor oil now being thrown away in the United States were recycled, it could save the country 1.3 million barrels of oil per day. Besides saving oil, recycling this waste can keep groundwater supplies significantly cleaner—one pint of oil poured down a storm drain or gutter can create a one-acre oil slick when it reaches the nearest stream, pond, or harbor.

Local recycling facilities, Environmental Protection Agency regional offices, and state solid waste offices are reliable sources of information about waste oil recycling in specific areas. Also, Valvoline and Jiffy Lube stations will accept motor oil for recycling, and Valvoline even has a hotline for finding their closest recycling location: 1-800-MOTOROIL. Many programs that accept waste oil for recycling charge 10 cents to 25 cents per gallon to cover their expenses.

Household Hazardous Waste (HHW)

Products considered to be household hazardous waste include used motor oil as well as most brands of paints and thinners, antifreeze, toilet bowl cleaners, drain openers, household bug sprays and other pesticides, nail polish, mothballs, batteries, and oven cleaners.

Every U.S. household currently produces approximately 21 pounds of HHW each year, most of which is disposed of improperly down drains or in landfills. But as communities have become increasingly aware of the problem, more programs are being developed to collect HHW for recycling, exchange, or proper handling and disposal.

In 1980, there were few, if any, programs for collecting HHW, but by 1993, over 7,400 one-time programs had been developed and run. In 1994, 30 states had set up 226 permanent HHW collection programs.

Latex paints, used oil, and antifreeze are the HHWs most commonly collected for recycling. In addition, most HHW programs stress the importance of not generating the waste in the first place by using nontoxic alternative products or buying only small quantities of the toxic products so that they will be used in their entirety, with no excess left for disposal.

Reliable sources of information on local HHW collection programs include regional Environmental Protection Agency offices, state health departments, and recycling facilities.

Batteries

There are two basic categories of batteries produced in the United States: automobile batteries and household batteries. Automobile batteries are lead-acid batteries, and there are numerous recycling programs throughout the country designed to handle this commodity, including national programs conducted by Sears and Wal-Mart. According to the EPA, the United States recycled 93.7 percent of its auto batteries in 1994, compared to 80 percent in 1989.

When car batteries are recycled, they are cracked open and the components are removed. The lead is recycled, and the sulfuric acid may either be reprocessed or sent to a hazardous waste disposal facility.

Household batteries are not being recycled on a widespread basis at this time. This category includes the following types of batteries: alkaline, carbon-zinc, nickel-cadmium, zinc-air, mercuric oxide, silver oxide, and lithium. Some of these are standard flashlight-type, cylindrical batteries; others are the button types used in cameras, hearing aids, watches, and calculators.

Although household batteries compose only about 0.005 percent by weight of the U.S. waste stream, they account for over 50 percent of the mercury and cadmium (both toxic metals) found in our trash. Recycling programs that accept batteries are often limited to the button types, from which mercury and silver may be recovered.

Most battery manufacturers are now offering rechargeable versions of household batteries in order to cut down on the number of batteries that need to be disposed of or recycled. This source reduction approach appears to be the one that most manufacturers prefer, rather than attempting to increase the recycling rate.

Rechargeable battery packs, most often found in cellular phones, most portable computers, power tools, and videotape recorders, are of the nickel-cadmium type. Several states have banned this type of battery from landfills, and as a result the Rechargeable Battery Recycling Corporation was formed by manufacturers, and a hotline was funded to assist consumers in recycling nickel-cadmium batteries: 1-800-8BATTERY.

Local sources of reliable information about battery recycling programs include Environmental Protection Agency regional offices, state offices of solid waste, HHW programs, and recycling facilities.

Aseptic Packaging and Polycoated Paper Packaging

Among the newer products entering the public recycling arena are polycoated paper packages, particularly milk cartons, and aseptic packaging in the form of juice boxes and other foods packaged in these cartons, such as tomatoes and soy milk. Manufacturers of these products, competing with the rest of the beverage packaging industry, are promoting their products as recyclable. The first drink box recycling programs were initiated by the aseptic industry in 1990. Today, postconsumer drink boxes are included in curbside collection programs serving close to 4 million households, according to industry information. At this time, however, there are few mills that are capable of hydropulping these types of containers and separating the plastic and aluminum components from the paper fibers.

Manufacturers of polycoated paper packages claim that recycling their products is both a boon to source reduction efforts and an energy-efficient process. Environmentalists, on the other hand, claim that the manufacturers are simply trying to sell the containers to stay in business and that this particular recycling process is far from energy efficient, especially because of the difficulty involved in collection.

Overall, aseptic packaging appears to rate higher as a source reduction tool than as a recyclable commodity, with its claim that most products packaged this way are typically 96 percent product and 4 percent packaging material by weight. More information on the recyclability of these products, as well as current recycling programs, can be obtained from the Aseptic Packaging Council's hotline: 1-800-277-8088.

White Goods

White goods include large appliances, such as water heaters, refrigerators, and clothes washers and dryers. In the past, these items were handled by scrap metals dealers who shredded and sold them. But when the EPA banned polychlorinated biphenyls, or PCBs, in 1979, scrap dealers were no longer so eager to collect white goods because the motors were likely to contain these deadly chemicals.

Currently, many trash hauling companies either pick up white goods for recycling for a fee (usually around $10) or offer programs at transfer stations that accept dropped-off white goods. Local recyclers, scrap metals dealers, or transfer stations should be consulted for specific information.

Laser Print Cartridges

In 1993, the United States used and discarded approximately 30 million laser cartridges. According to Environmental Protection Agency estimates, 99 million ink-jet cartridges will be consumed in 1997. But a cartridge can be cleaned and refilled at a significant saving over the cost of a new one, and over the last few years the business of recycling these cartridges has grown significantly. Refilling rates are expected to reach 22 percent in 1998.

Many such cartridges now come in packaging that includes manufacturer's instructions on how to recycle their product (there is also a company, Nu·Kote, that sells a refillable cartridge system; the print head is a one-time purchase, while the ink cartridges are refillable and can be sent back to the manufacturer).

A local computer company or office supply store that sells these cartridges can offer information on recycling them. The trade association for these products is the Imaging Products Remanufacturers Association in Silver Springs, Maryland (phone: 301-588-6777).

Old Sneakers

Nike has begun a new program that recycles old fitness footwear through their *Reuse-a-Shoe*™ Program. The rubber soles are recovered from the shoes, ground up, and used to create surfacing for basketball courts, running tracks, and playgrounds across the nation. Over 2 million shoes have been recycled through this program. For more information, call 1-800-352-6453.

Paper Sludge

One company, Petruzzo Products of New Jersey, is currently taking paper sludge from a local mill and turning it into cat litter that will absorb four to five times as much odor as traditional clay cat litter.

Dry Wall

Canagro Agricultural Products, Ltd., which manufactures organic fertilizers in Elmira, Ontario, began successfully recycling drywall gypsum into fertilizer, cat litter, and livestock bedding in 1989.

Single-Use Cameras

In 1995, the Eastman Kodak Company recycled its single-use cameras at a rate of 77 percent (even higher than the recycling rate of aluminum for that same year). As of 1996, more than 70 million of these cameras have been recycled and/or reused through Kodak's program. Since the beginning of its recycling program, Kodak has diverted over 9.8 million pounds of material from landfills (the equivalent of 774 tractor-trailer loads).

Kodak Fun Saver™ 35 single-use cameras are designed for recycling, since the consumer must take the entire camera into the photo processor in order to get the film developed. Each new camera sold contains recycled plastic and/or metal parts, with a new lens and new battery.

More information on Kodak's recycling programs can be obtained by calling the Kodak Information Center at 1-800-242-2424.

Film and Film Containers

Kodak also has a film container recycling program, initiated in 1992. Approximately 26 million pounds of waste have been diverted from landfills, in the way of plastic and metal materials from 1 billion rolls of 35 mm film, since the beginning of this program.

Additionally, Kodak recycles 35 mm film materials from all manufacturers and recycles or reuses the materials, including the polystyrene spool and the steel on either end of the spool, the steel cartridge itself, and the polyethylene container and lid.

More information on Kodak's recycling programs can be obtained by calling the Kodak Information Center at 1-800-242-2424.

Toilets

In Santa Barbara County, California, old porcelain toilets received at landfills are crushed and reused in roadway paving projects.

Telephone Directories

Until fairly recently, hot glue bindings, poor-quality groundwood pages, and a different grade of cover stock have combined to make phone books a difficult recycling proposition. However, due to pressure from consumers, most manufacturers of telephone directories are now offering recycling programs for their old books and are producing new directories from recycled paper. One program in New

York composts the shredded books with sawdust and beer sludge from Anheuser-Busch. In other programs, directory manufacturers combine forces with local trash haulers to provide drop-off locations for the directories, which will be recycled into new paper after the bindings have been removed.

Some states are now legislating requirements for the recycling of phone directories as well as mandating a recycled-content percentage for new directories.

Compact Discs (CDs)

The largest manufacturer of CDs in the United States, Digital Audio Disk Corporation, is currently investigating a technology to recycle CDs in its plant in Terre Haute, Indiana. There, discs that have been rejected from the factory are recycled into the trays in which CDs are packaged.

Personal Computers

Several companies are accepting used personal computers for refurbishing and resale. For more information, consult local computer stores. (One such company is Recompute, a mail-order business based in Austin, Texas.)

Sources

The Aluminum Association. *Aluminum Recycling: America's Environmental Success Story.* Washington, DC: The Aluminum Association, 1990.

American Forest & Paper Association. *PaperMatcher™—A Directory of Paper Recycling Mills.* 4th ed. Washington, DC: American Forest & Paper Association, 1996.

———. *Recovered Paper Statistical Highlights.* 1995 ed. Washington, DC: American Forest & Paper Association, 1996.

American Iron and Steel Institute. *Today's Steel Is Tomorrow's Steel.* Internet: http://www.steel.org.

American Paper Institute, Inc. *Paper Recycling and Its Role in Solid Waste Management.* New York: American Paper Institute, 1990.

———. *12 Facts about Waste Paper Recycling.* New York: American Paper Institute, 1990.

American Plastics Council. *What Industry Is Doing.* Washington, DC: American Plastics Council, 1996.

Apotheker, Steve. "Glass Containers: How Recyclable Will They Be in the 1990s?" *Resource Recycling* (June 1991): 25–32.
———. "Office Paper Recycling: Collection Trends." *Resource Recycling* (November 1991): 44–52.
Aseptic Packaging Council. *Drink Box Recycling*. Washington, DC: Aseptic Packaging Council, 1996.
———. *Packaging for a Healthy Lifestyle*. Washington, DC: Aseptic Packaging Council, 1996.
Ball, Doug. "Recycled Phone Books Find Home in Brewery Sludge." *BioCycle* 31 (November 1991): 62.
Calem, Robert E. "Recycling the Proliferating Disk." *New York Times* (12 January 1992).
The Copper Page. *Copper—The World's Most Reusable Resource*. Internet: http://www.copper.org.
The Council for Solid Waste Solutions. *The Facts about Plastics*. Washington, DC: Society of the Plastics Industry, 1990.
Davis, Alan, and Susan Kinsella. "Recycled Paper: Exploding the Myths." *GARBAGE* 2 (May/June 1990): 48–54.
The EarthWorks Group. *The Recycler's Handbook*. Berkeley, CA: EarthWorks Press, 1990.
———. *50 Simple Things You Can Do to Save the Earth*. Berkeley, CA: EarthWorks Press, 1989.
———. *50 Simple Things Your Business Can Do to Save the Earth*. Berkeley, CA: EarthWorks Press, 1991.
Emm, Joseph. "Hazardous Household Waste." *Home* (October 1996): 140–144.
Environmental Action, eds. "Recycling Plastics: A Forum." *Environmental Action* (July/August 1988): 21–25.
Environmental Defense Fund. *Recycle: It's the Everyday Way to Save the World*. Washington, DC: Environmental Defense Fund, 1990.
Erkenswick, Jane L. "Office Paper Recycling: A Look at the Ledger Grades." *Resource Recycling* (November 1991): 64–68.
Franklin Associates, Ltd. *Waste Management and Reduction Trends in the Polystyrene Industry, 1974–1994*. Prairie Village, KS: Franklin Associates, 1996.
Glass Packaging Institute. *Americans Prefer Glass*. Washington, DC: Glass Packaging Institute, 1990.
———. *Package to Package and Glass to Glass: The Perfect Closed Cycle System*. Washington, DC: Glass Packaging Institute, 1990.
Institute of Scrap Recycling Industries. *Recycling Nonferrous Scrap Metals*. Washington, DC: Institute of Scrap Recycling Industries, 1990.

———. *Recycling Paper.* Washington, DC: Institute of Scrap Recycling Industries, 1990.

———. *Recycling Scrap Iron and Steel.* Washington, DC: Institute of Scrap Recycling Industries, 1990.

———. *Scrap Specifications Circular/Guidelines for Paper Stock: PS-90, Domestic Transactions.* Washington, DC: Institute of Scrap Recycling Industries, 1990.

———. *Value.* Washington, DC: Institute of Scrap Recycling Industries, 1996.

International Tire & Rubber Association. *Tire & Rubber Recycling Fact Sheet.* Louisville, KY: International Tire & Rubber Association, 1996. Internet: http://www.itra.com.

Logsdon, Gene. "Agony and Ecstasy of Tire Recycling." *BioCycle* 30 (July 1990): 44–85.

Luoma, Jon R. "Trash Can Realities." *Audubon* 92 (March 1990): 86–97.

McEntee, Ken. "Office Paper Recycling: Boom or Bust?" *Resource Recycling* (November 1991): 58–63.

National Association for Plastic Container Recovery. *Recycling PET: A Guidebook for Community Programs.* Charlotte, NC: National Association for Plastic Container Recovery, 1989.

National Recycling Coalition. *Recycled Paper Facts and Figures* (fact sheet). Washington, DC: National Recycling Coalition, 1990.

National Soft Drink Association. "Closing the Loop." *The Soft Drink Recycler* (Spring 1991): 1–2.

National Solid Wastes Management Association. *The Future of Newspaper Recycling.* Washington, DC: National Solid Wastes Management Association, 1990.

———. *Recycling Solid Waste.* Washington, DC: National Solid Wastes Management Association, 1990.

Peters, Dale T. "Trends in U.S. Copper Alloy Scrap and Effects of Product Shifts." World Conference on Copper Recycling. Brussels, Belgium, 4 March 1997.

Porter, J. Winston. "Let's Go Easy on Recycling Plastics." *Philadelphia Inquirer* (6 August 1991).

Powell, Jerry, and Ken McEntee. "Office Paper Recycling: Existing and Emerging Markets." *Resource Recycling* (November 1991): 53–57.

Rembert, Tracey C. "Bananas Save Trees." *E Magazine* (January/February 1997): 23.

———. "Dam Those Tires!" *E Magazine* (January/February 1997): 21-22.

Reutlinger, Nancy, and Dan de Grassi. "Household Battery Recycling: Numerous Obstacles, Few Solutions." *Resource Recycling* (April 1991): 24–29.

Steel Recycling Institute. *Recyclable Steel Cans: An Integral Part of Your Curbside Recycling Program.* Pittsburgh, PA: Steel Recycling Institute, 1990.

Thompson, Claudia G. *Recycled Papers: The Essential Guide.* Cambridge, MA: MIT Press, 1992.

U.S. Environmental Protection Agency. *Characterization of Municipal Solid Waste in the United States: 1990 Update, Executive Summary.* Washington, DC: U.S. Environmental Protection Agency, 1990. EPA/530-SW-90-042.

———. MSW Factbook, Ver. 3.0. Washington, DC: U.S. Environmental Protection Agency, 1996. EPA/530-C-96-001.

———. *Summary of Markets for Scrap Tires.* Washington, DC: U.S. Environmental Protection Agency, 1991. EPA/530-SW-90-074B

———. *Yard Waste Composting: A Study of Eight Programs.* Washington, DC: U.S. Environmental Protection Agency, 1989.

Wirka, Jeanne. "A Plastics Packaging Primer." *Environmental Action* (July/August 1988): 17–20.

State Laws and Regulations

Solid waste and recycling legislation has become a volatile subject ever since the first bottle bill—requiring retailers to collect empty beverage bottles from consumers and refund a cash deposit—was passed in Oregon in 1971. Such laws are generally welcomed by the public but abhorred by industrial interests, whose profit margins are affected by any form of mandatory recycling requirements.

The state of the PET (polyethylene terephthalate) recycling market provides a good case in point. In 1996, after most of the state laws that required PET recycling were either substantially weakened or repealed entirely, the recycling industry watched PET prices plummet and markets for this commodity all but disappear. It was estimated that only about 12 of the 32 major PET reclaimers were buying the resin as of December 1996.

Industry observers have criticized the plastics industry, claiming that the only reason PET recycling had occurred at such strong rates in the past was because of legislative requirements, and when those disappeared there was no reason to continue doing anything that challenged the bottom line, regardless of the environmental ethics involved. As long as it is more expensive to use recycled plastic resin instead of virgin resin, as is currently the case, and there is no legislation to supply economic incentives to do otherwise, recycling experts agree that PET marketing problems will continue.

But the PET issue is only one example of what is going on in the world of recycling when it comes to state regulation. The connection between economics and the success of recycling programs is one that is unlikely to break apart. As noble as environmental ethics may be, most experts in the solid waste field agree that the majority of businesses will not recycle, nor will they buy or produce recycled products, unless it provides them with a profit or, at the very least, does not detract from their profit margins.

As a result of these factors, many states are presently considering reassessing their recycled-content mandates. Wisconsin and Minnesota, specifically, have registered their intentions to address packaging initiatives in 1997 with regard to recycled-content mandates, and California may strengthen its recycled-content laws, especially with respect to PET.

Meanwhile, during the 1997 legislative session, several state legislatures are expected to examine bottle deposit legislation. As with other types of recycling legislation, bottle deposits are much more popular with the public than they are with retailers and the beverage industry. Massachusetts, Michigan, Oregon, and New York are expected to consider expanded deposits (raising the amount of money required as a deposit to cover handling fees and to increase revenues), while the state of Georgia is considering initiating a deposit law for the first time.

At the moment, the ten states that do have bottle bills have had them for years, and industrial interests have successfully thwarted such legislation in most of the other states where it has been proposed by citing the difficulties involved for small store owners and the fact that these bills only address a relatively small portion of the entire waste stream. On the other hand, bottle bill proponents repeatedly point out that states with such bills recycle twice as many containers as those without them. Georgia is the first state to consider implementing such a law in several years (although expanded deposit referendums have come up regularly in states that already have bottle deposits).

Solid waste management issues have given rise to numerous state laws regarding recycling, most of which fall into one of the following four categories: (1) container deposit laws, (2) landfill bans, (3) recycled-content mandates, and (4) recycling materials requirements that define which materials have to be recycled by state residents and businesses. Many of these laws are described in this chapter, but because legislation varies widely in its force and language, and because new legislation and/or amendments to existing statutes are introduced each year, the reader is advised to examine actual statutes in their entirety for legal purposes.

Figures 5.1 through 5.9 and Tables 5.1 through 5.4 summarize much of the recycling legislation in the United States and are provided courtesy of the National Solid Wastes Management Association. More detailed examinations of recycling legislation, state by state, follow. This information is derived from the WESTLAW database as well as compiled through research of actual state legislation. Federal laws are discussed in Chapter 1. State recycling rates as of April 1996 were compiled from *BioCycle* magazine, "The State of Garbage in America" (April 1996), as referenced on the Natural Resources Defense Council's home page on the Internet (see Chapter 8).

Figure 5.1 States Having Comprehensive Recycling Laws as of 31 December 1990

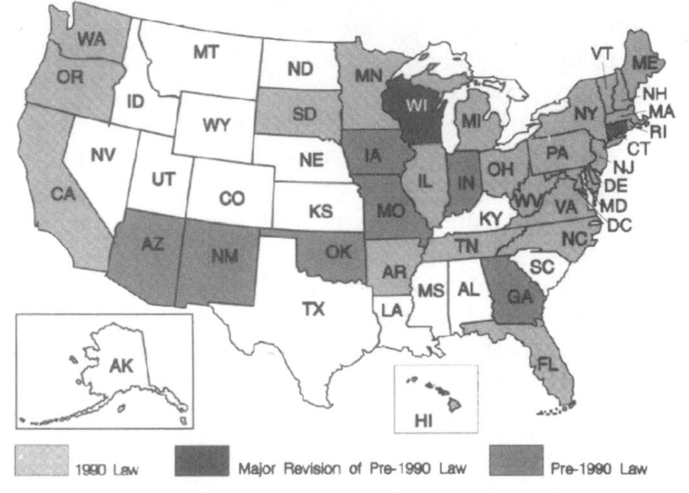

Note: Comprehensive recycling laws require detailed statewide recycling plans and/or separation of recyclables and contain at least one other provision to stimulate recycling

Source: Adapted from National Solid Wastes Management Association, *Recycling in the States 1990 Review,* Washington, DC: National Solid Wastes Management Association, 1990.

Figure 5.2 Recycling Grants and Loans, by State

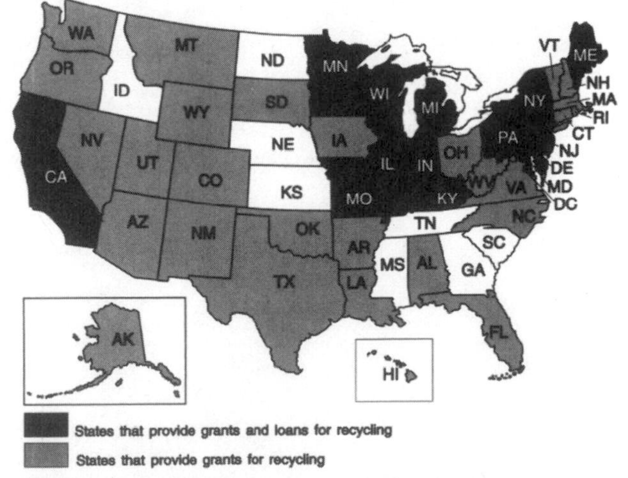

Source: Printed from the MSW Factbook, Ver. 3.0, Office of Solid Waste, U.S. Environmental Protection Agency, Washington, DC..

Figure 5.3 Recycled Content Mandates, by State

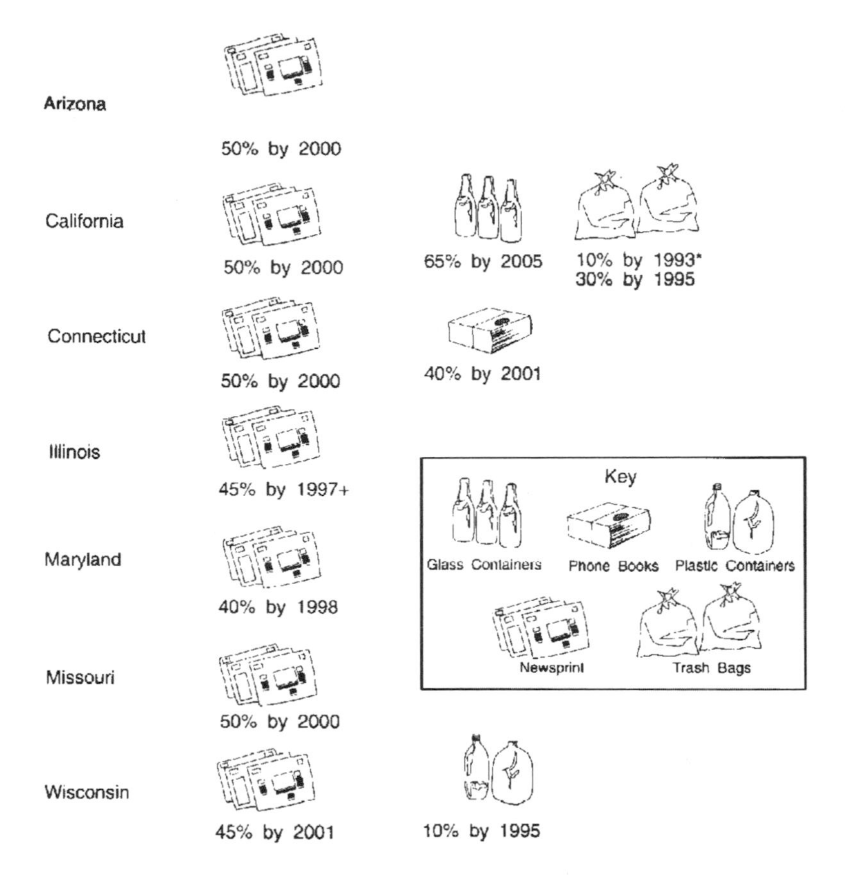

Note: These percentages represent the total amount of recycled material that must be used. The required ratio of postconsumer to industrial scrap in recycled material varies in each state.

* The 10 percent goal applies to bags 1.0 mil thick; 30 percent goal applies to bags .75 mil thick.

\+ The 45 percent goal by 1997 is a voluntary goal; the mandatory goal is 28 percent by 1993.

Source: Adapted from National Solid Wastes Management Association, *Recycling in the States 1990 Review,* Washington, DC: National Solid Wastes Management Association, 1990.

Figure 5.4 States with Bottle Deposit Rules

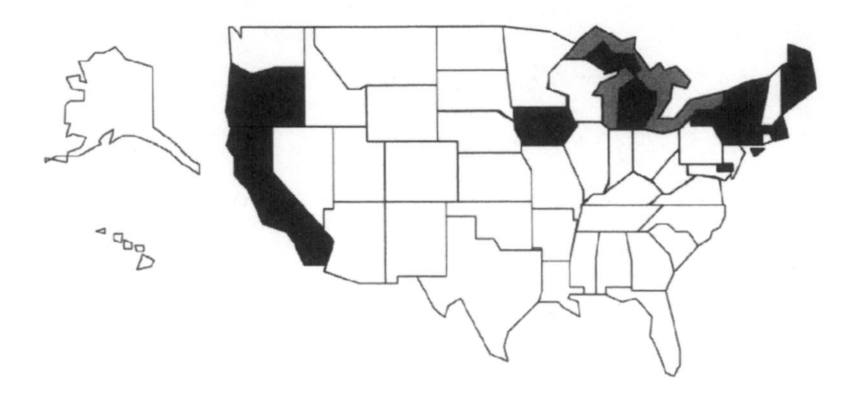

Source: Printed from the MSW Factbook, Ver. 3.0, Office of Solid Waste, U.S. Environmental Protection Agency, Washington, DC., 1996.

Figure 5.5 States with Yard Waste Bans

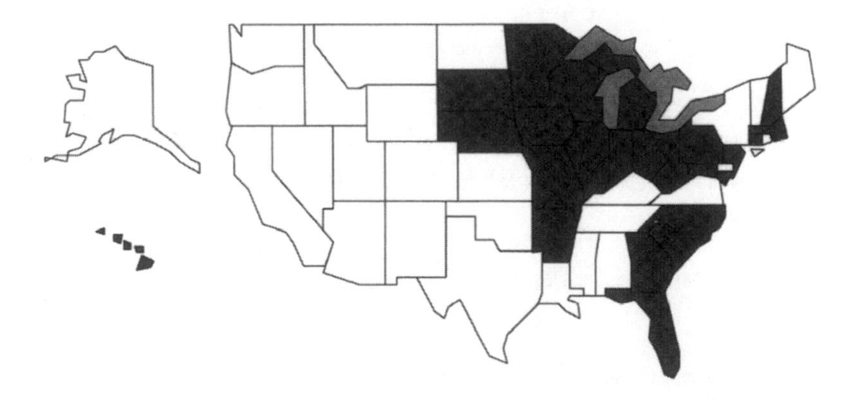

Source: BioCycle Magazine. Printed from the MSW Factbook, Ver. 3.0, Office of Solid Waste, U.S. Environmental Protection Agency, Washington, DC., 1996.

Figure 5.6 Number of Yard Waste Composting Programs

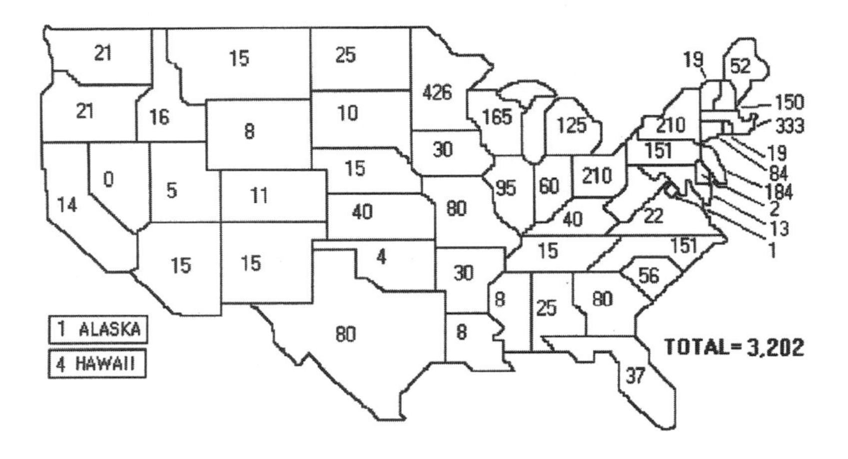

Source: BioCycle Magazine. Printed from the MSW Factbook, Ver. 3.0, Office of Solid Waste, U.S. Environmental Protection Agency, Washington, DC., 1996.

Figure 5.7 Permanent Household Hazardous Waste Collection Programs

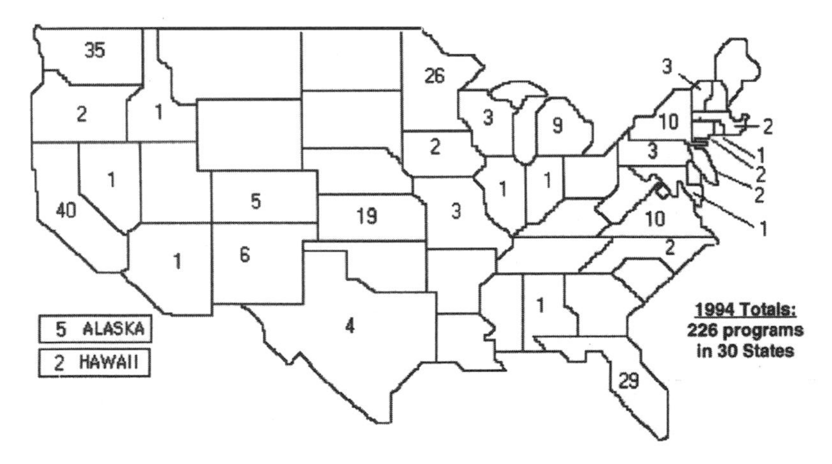

Source: The Waste Watch Center, Andover, MA, 1995. Printed from the MSW Factbook, Ver. 3.0, Office of Solid Waste, U.S. Environmental Protection Agency, Washington, DC., 1996.

Figure 5.8 State Recycling Rates

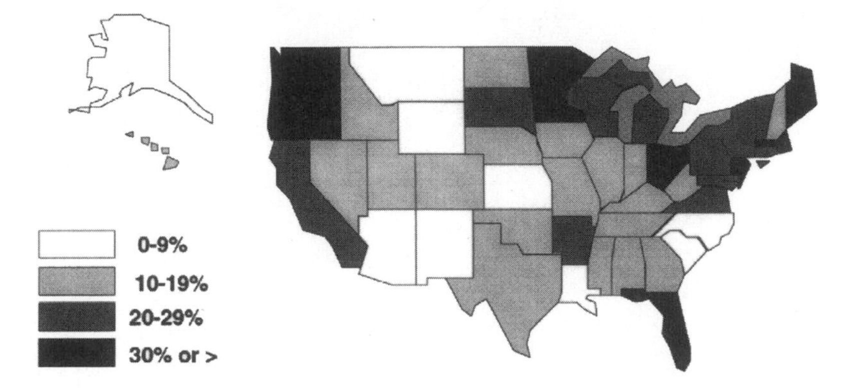

Source: BioCycle Magazine. Printed from the MSW Factbook, Ver. 3.0, Office of Solid Waste, U.S. Environmental Protection Agency, Washington, DC., 1996.

Figure 5.9 Recycling Rates of Selected Materials, 1994

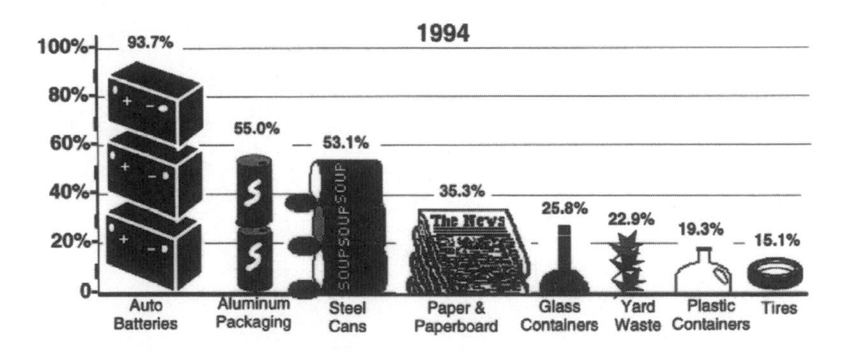

Source: EPA Waste Characterization Report, Franklin Assoc., 1995. Printed from the MSW Factbook, Ver. 3.0, Office of Solid Waste, U.S. Environmental Protection Agency, Washington, DC., 1996.

Table 5.1 Types of State Recycling Laws through 1990

State	Year Enacted	Type of Plan
Arizona	1990	Opportunity to recycle only
Arkansas	1989	Recycling plans only
California	1989	Mandatory goals*
Connecticut	1990	Source separation
Delaware	1990	Opportunity to recycle only
District of Columbia	1989	Source separation
Florida	1988	Mandatory goals*; community separation
Georgia	1990	Community separation
Hawaii	1988	Opportunity to recycle only
Illinois	1988	Community separation
Indiana	1990	Recycling plans only
Iowa	1989	Recycling plans only
Maine	1989	Source separation+
Maryland	1988	Mandatory goals*; community separation
Massachusetts	1987	Recycling plans only
Michigan	1988	Recycling plans only
Minnesota	1989	Mandatory goals*; community separation
Missouri	1990	Recycling plans only
New Hampshire	1988	Recycling plans only
New Jersey	1987	Mandatory goals*; source separation
New Mexico	1990	Recycling plans only
New York	1988	Source separation
North Carolina	1989	Community separation
Ohio	1988	Mandatory goals*; community separation
Oklahoma	1990	Recycling plans only
Oregon	1983	Opportunity to recycle only
Pennsylvania	1988	Source separation
Rhode Island	1986	Mandatory goals*; source separation
Tennessee	1989	Recycling plans only
Vermont	1987	Recycling plans only
Virginia	1989	Mandatory goals*
Washington	1989	Community separation
West Virginia	1989	Recycling plans only
Wisconsin	1990	Community separation

* "Mandatory" can be subject to different interpretations.

\+ For office use only.

Source: Adapted from National Solid Wastes Management Association, *Recycling in the States 1990 Reveiw,* Washington, DC: NSWMA, 1990.

Table 5.2 States with Source Separation Requirements, 1990

State	Who Must Separate	Materials
District of Columbia	Residences, businesses	Glass containers, metal cans, newspapers, yard waste*, paper+
Connecticut	All generators	Glass and metal food containers, newspaper, cardboard, office paper, used oil, car batteries, nickel-cadmium batteries, leaves, scrap metal
Maine	Businesses	Office paper, corrugated cardboard
New Jersey	Residences, businesses, institutions	Three materials§ and leaves
New York	Residences, businesses, institutions	Paper, glass, metal cans, plastic containers, yard waste**
Pennsylvania	Residences	Three materials§ and leaves
	Businesses, institutions	High-grade office paper, corrugated cardboard, aluminum cans++
Rhode Island	Residences, businesses	Glass food and beverage containers, newpaper, tin and steel cans, aluminum, some plastics, large appliances
	Businesses	Cardboard, office paper
Washington	Urban areas	Materials to be determined in local plans

Note: This chart does not include separation requirements that only apply to state government agencies and institutions.

* Residential separation only
\+ Separation by offices only
§ Municipalities choose three materials to be recycled from a state list.
** If "economically feasible."
++ Municipalities may require businesses to separate additional recyclables.

Source: Adapted from National Solid Wastes Management Association, *Recycling in the States 1990 Review,* Washington, DC: NSWMA, 1990.

Table 5.3 States with Disposal Bans, 1990

State	Lead-Acid Batteries	Yard Waste	Unprocessed Tires	Used Oil	Large Appliances	Other
California	•					
Connecticut	•	*		•		A
District of Columbia				•		
Florida	•	+	•	•	•	B
Georgia	•					
Hawaii	•					
Illinois	•	•	•			
Iowa	•	•	•	•		C
Kansas	•		•			
Kentucky	•					
Louisiana	•		•	•	•	
Maine	•					
Massachusetts	•	•	•	•	•	D
Michigan	•	•		•		
Minnesota	•	•	•	•	•	E
Missouri	•	•	•	•	•	
New Hampshire	•					
New Jersey		*				
New York	•					
North Carolina	•	+	•	•	•	
Ohio	•	•	•			
Oregon	•		•			F
Pennsylvania	•	*				
Rhode Island			§			G
Tennessee	•					
Vermont	•		•	•	•	
Virginia	•					
Washington	•					
Wisconsin	•	•	**	**	•	H
Wyoming	•					

A Nickel-cadmium batteries.
B Construction and demolition debris
C Nondegradable grocery bags; beverage containers returned to wholesalers through the state's mandatory deposit law.
D Aluminum, glass, and metal containers, single polymer plastics, and recyclable paper.
E Dry-cell batteries that contain mercuric oxide electrodes, nickel-cadmium, or sealed lead-acid. Mixed unprocessed waste in metro area.
F Recyclable material that has already been separated.
G Loads of commercial waste containing more than 20 percent recyclables.
H Nondegradable yard waste bags plus aluminum, plastic, steel, and glass containers, cardboard foam polystyrene packaging, magazines, newspaper, and office paper are banned from disposal unless municipalities are certified as having an "effective" source separation program.
* Yard waste disposal bans only apply to leaves.
+ Ban applies to lined landfills only.
§ Banned only from incinerators.
** Can be incinerated with energy recovery.

Source: Adapted from National Solid Wastes Management Association, *Recycling in the States 1990 Review,* Washington, DC: National Solid Wastes Management Association, 1990.

Table 5.4 Tax Incentives for Recycling, by State

California—Banks and corporations may take 40 percent tax credit for the cost of equipment used to manufacture recycled products. Development bonds for manufacturing products with recycled materials.

Colorado—Individual and corporate income tax credits for investments in plastics recycling technology.

Florida—Sales tax exemption on recycling machinery purchased after 1 July 1988. Tax incentives to encourage affordable transportation of recycled goods from collection points to sites for processing and disposal.

Illinois—Sales tax exemptions for manufacturing equipment.

Indiana—Property tax exemptions for buildings, equipment, and land involved in converting waste into new products.

Iowa—Sales tax exemptions.

Kentucky—Property tax exemptions.

Maine—Business tax credits equal to 30 percent of cost of recycling equipment and machinery. Subsidies to municipalities for scrap metal transportation costs. Taxpayers are also allowed a credit equal to $5 per ton of wood waste from lumber production that is incinerated for fuel or to generate energy. The total credit may not exceed 50 percent of the tax liability.

Maryland—From their state income taxes, individuals and corporations can deduct 100 percent of expenses incurred to convert a furnace to burn used oil or to buy and install equipment to recycled used freon.

New Jersey—Businesses may take a 50 percent investment credit for recycling vehicles and machinery, 6 percent sales tax exemption on purchases of recycling equipment.

North Carolina—Industrial and corporate income tax credits and exemptions for equipment and facilities.

Oregon—Individuals and corporations receive income tax credits for capital investment in recycling equipment and facilities. Special tax credits are available for equipment, property, or machinery necessary to collect, transport, or process reclaimed plastic.

Texas—Sludge recycling corporations are eligible for franchise tax exemptions.

Virginia—Individuals and corporations may take a tax credit worth 10 percent of the purchase price of any machinery and equipment for processing recyclable materials. The credit also applies to manufacturing plants that use recycled products.

Washington—Motor vehicles are exempt from rate regulation when transporting recovered materials from collection to reprocessing facilities and manufacturers. Tires and certain other hard-to-dispose materials are exempt from portions of sales and use taxes.

Wisconsin—Sales tax exemptions for waste reduction and recycling equipment and facilities; business property tax exemptions for some equipment.

Source: Adapted from National Solid Wastes Management Association, *Recycling in the States 1990 Review,* Washington, DC: NSWMA, 1990.

Alabama

- As of April 1996, Alabama's recycling rate was 15 percent.
- On 19 April 1990, Alabama mandated that, within 180 days, the Department of Environmental Management would "develop and implement a model program for the reduction and recycling of solid wastes within its own operations." This program was to include but was not limited to office papers, cardboard, yard waste, and "other materials produced by the state for which recycling markets exist or may be developed."
- Within one year after 19 April 1990, all state departments and agencies, as well as all public schools, were required to implement similar programs based on the model provided by the Department of Environmental Management. The Alabama Environmental Management Act states that the products to be recycled must include, at a minimum, both high-grade paper and corrugated paper. The act also authorizes state agencies and public schools to enter into contracts with private and nonprofit organizations to manage and sell or donate recyclable materials, as long as these materials are in fact "substantially recycled." Furthermore, any proceeds from the sale of recycled materials must be deposited back into the state treasury and credited to the agency that originally generated the recyclables.

SOURCE: State Code of Alabama 1975: Title 22, Subtitle 1, Chapter 22B (Recycling by State Agencies), §22-22B-1 through §22-22B-5; Alabama Law, Act No. 89-824 and 90-564.

Alaska

- As of April 1996, Alaska's recycling rate was 6 percent.
- When Alaska created and empowered its Department of Environmental Conservation in 1990, the department was directed to coordinate state efforts between public and private organizations with regard to recycling and litter control. Waste source reduction and the recycling of waste were established as the two top priorities when dealing with solid and hazardous waste, and a community solid waste management planning grant account was created within the state's general fund.
- In 1980, all state agencies were required to recycle at least paper, glass, and cans "to the greatest extent practicable."

- Beginning on 1 October 1981, all detachable, metal pull-tabs on nonglass beverage containers were prohibited. And on 1 January 1985, plastic six-pack rings and similar packaging devices were outlawed, unless they were demonstrably degradable and bore a distinguishing mark that designated them as such. Then, in 1990, a requirement was added stipulating that all plastic bottles or rigid plastic containers sold within the state must bear the plastics code as designated by the Society of the Plastics Industry (see Chapter 4 for more information on this code).
- In 1991, the Department of Environmental Conservation was empowered to establish a waste reduction and recycling awards program in consultation with the Department of Education. The program would provide, as funds were available, award grants for public schools to recognize their efforts to reduce and recycle waste they had generated.
- By 1995, the state added some incentives for buying recycled materials: a 5 percent procurement preference for recycled-content purchases in general, and a 10 percent price preference for buying retreaded tires.

SOURCE: Alaska State Statutes 1962–1991: Title 46, Chapter 6 (Recycling and Reduction of Litter), §46.06.010, §46.06.021, §46.11.060, §46.06.090(a) and (b); §46.06.095, §46.11.070. U.S. Environmental Protection Agency, MSW Factbook, Ver. 3.0, 1996.

Arizona

- As of April 1996, Arizona's recycling rate was 10 percent.
- On 28 June 1990, Governor Rose Mofford approved the Arizona Solid Waste Recycling Act (effective 27 September 1990) requiring that residents be provided with "an opportunity to engage in recycling and waste reduction."
- The state mandates that all motor vehicle tires shall be recycled and requires all sellers of tires to post written notices containing the following language: "It is unlawful to throw away a motor vehicle tire. Recycle all used tires." Retailers must accept scrap tires if any new or retreaded tires are sold, and when any new tire is purchased, an additional fee for environmental protection is charged.
- The disposal of automotive lead-acid batteries in landfills or incinerators is prohibited. These batteries must be collected and recycled.

- The Department of Environmental Quality is empowered by the state to administer, encourage, and oversee the state's recycling programs. These include a public education program, run in consultation with the State Board of Education, that addresses all facets of recycling, from source reduction to the promotion of buying recycled products.
- As of 1 July 1991, all consumers of newsprint within the state were required to use at least 25 percent recycled content in their papers (if the cost of the recycled newsprint is within 5 percent of the cost of virgin material newsprint). Further goals were specified with regard to recycled-content newsprint: 30 percent on 1 January 1994, 35 percent on 1 January 1996, 40 percent on 1 January 1998, and 50 percent on 1 January 2000.
- As of 1 July 1991, it was also mandated that all plastic bottles and rigid plastic containers must bear the recycling symbols defined by the Society of the Plastics Industry to designate the resin used in their manufacture.
- A state recycling fund was established in 1991 to help finance a variety of interests revolving around recycling, including research, education, market development, and source reduction. It specified that at least 40 percent of the monies be directed to grants and at least 20 percent to public education and information. No more than 5 percent may be allocated for collection and administration of the monies and no more than another 5 percent for the administration of the article (§49-837).
- In December 1992, a 10 percent tax credit for anyone purchasing recycling equipment was added to state law. At the same time, a state Recycled Materials Market Development program was created, and the recycling fund was continued.

SOURCE: Arizona Revised Statutes: Title 9 (Cities and Towns), §9-500.07; Title 11 (Counties), §11-269; Title 44 (Trade and Commerce), Chapter 9, Article 8 (Waste Tire Disposal), §44-1302, Article 9 (Lead-Acid Batteries), §44-1322; Title 49 (Arizona Recycling Program), Article 8, Chapter 4, §49-833, §49-834, §49-835, §49-837. State of Arizona Senate Bill 1287, amending sections of Title 41, 43, and 44, adding sections 44-1644 and 44-1645. The Arizona Solid Waste Recycling Act, Chapter 378, House Bill 2574.

Arkansas

- As of April 1996, Arkansas' recycling rate was 25 percent.
- Arkansas mandated that all state agencies submit "a solid waste management plan which proposes the establishment

of recycling programs and facilities" by 1 July 1991. Subsequently, the state created a Solid Waste Management and Recycling Fund, administered by the Department of Pollution Control and Ecology.

- State law requires that the public is assured an "opportunity to recycle." Specifically, all residents must have access to "curbside pick-up or collection centers for recyclable materials at sites that are convenient for persons to use." Further, as of 1 July 1993, "at least one (1) recyclable materials collection center shall be available in each county of a district unless the commission grants the district an exemption." Exemptions are possible if such a collection facility exists in a nearby county and is shown to serve both areas adequately.
- A State Marketing Board for Recyclables has been established to promote all aspects of recycling and stimulate markets for recyclable materials. Additionally, all state agencies and schools have been required to institute source separation recycling programs, and they are encouraged to give purchasing preference to products made from recycled materials or products that may be readily recycled or reused.
- The state purchasing agent has been directed to issue specifications for recycled paper content on all types of papers. Also, purchasing goals dealing with the percentage of recycled paper products to be used within state agencies were delineated as follows: 10 percent in fiscal year 1991, 25 percent in fiscal year 1992, 45 percent in fiscal year 1993, and 60 percent by calendar year 2000.
- When paper products are put out to bid, a 10 percent difference in price is permitted in recycled versus virgin paper goods. Also, an additional 1 percent preference is allowed for "products containing the largest amount of postconsumer materials recovered within the State of Arkansas."
- The state grants a variable tax credit for the purchase of waste reduction, reuse, or recycling equipment that is used to process products made of at least 50 percent recovered waste materials, 10 percent of which includes postconsumer waste.
- Beginning in 1993, retailers of lead-acid batteries were required to accept used batteries for recycling when selling new ones. In fact, customers are to be charged $10 when purchasing a new lead-acid battery; if they return a used battery within 30 days, they receive the surcharge back. Lead-acid batteries may no longer be disposed of in landfills (specifica-

tions are made for where these batteries can be dropped off for safe disposal or recycling).

- Also in 1993, a fee system was imposed for the importation of any waste tires from out-of-state. Additionally, a variable state income tax credit was established for the purchase of "waste reduction, reuse, or recycling equipment."
- In 1995, a new law passed making it illegal to place yard waste (grass clippings, leaves, and shrubbery trimmings) in a solid waste management facility, effectively making it necessary to both source reduce and compost these wastes. At the same time, an act passed to "provide that regional solid waste management boards may adopt more restrictive standards for the location, design, construction, and maintenance of solid waste disposal sites."
- Also in 1995, an Arkansas Newspaper Recycling Advisory Committee was established as part of the Department of Pollution Control and Ecology.

SOURCE: Arkansas Code of 1987: Title 8 (Environmental Law), Chapter 6 (Disposal of Solid Wastes and Other Refuse), Subchapter 6 (Solid Waste Management Recycling Fund Act), §8-6-604, §8-6-605, Subchapter 7, §8-6-720, §8-9-201, §8-9-203, §8-9-204, §19-5-961, §19-11-260, §26-51-506. State Senate Bills 421 (1993), 431 (1995), 449 (1993), 1634 (1995), 1931 (1995), amending sections of Title 8.

California

- As of April 1996, California's recycling rate was 25 percent.
- All state school districts and universities are encouraged to establish paper recycling programs and to cooperate with each other in setting up such programs. State educational agencies are also encouraged to purchase recycled paper "if the supplier of recycled paper offers the paper at a cost which does not exceed by more than 5 percent the lowest offer of nonrecycled paper of comparable quality." Moreover, state law requires that the recycled paper that is purchased should have the highest percentage of postconsumer waste, all other qualities being equal. Additionally, the agencies are asked to make all reasonable efforts to eliminate potential paper contaminating products from their purchasing programs.
- Purchase preference for recycled paper by universities and state agencies is encouraged, and goals were set for purchasing

such products: By 1 January 1992, at least 35 percent of the total dollar amount of paper products purchased was to be recycled; by 1 January 1994, at least 40 percent; and by 1 January 1996, at least 50 percent. The regents of the University of California are required to submit an annual report to the state legislature, the governor, and the California Integrated Waste Management Board that details the percentage of the total dollar amount spent that is devoted to recycled paper products.

- The state requires written certification of the percentage of secondary and postconsumer waste in all recycled paper products. Where public contracts are concerned, all contractors must certify in writing the minimum or exact percentages of both postconsumer and secondary waste contained in their products.
- Persons contracted by the state are required to use recycled paper products to the maximum extent deemed "economically feasible." Any company bidding on a printing project for any state or local agency must specify the percentage of recycled materials in their recycled paper products. The following purchasing goals for high-grade recycled papers by state agencies were established: 30 percent by 1 January 1994, 35 percent on and after 1 January 1997, and 40 percent on and after 1 January 2000.
- All state agencies and the legislature were required to have waste paper collection and recycling programs in operation by 1 June 1991; these must include educational programs for all personnel and must provide for the collection of office paper, corrugated cardboard, newsprint, beverage containers, waste oil, and "any other material at the discretion of the director," in consultation with the California Integrated Waste Management Board. Constant reevaluation of the program is specified, and there is a provision for discontinuing certain products if the economics of the program prove ineffective.
- Any revenues realized from these recycling programs are to be used according to the following priorities, in the order given: to offset recycling program costs, to offset the higher prices of purchasing recycled paper products, and, lastly, to augment California's general fund.
- All procuring agencies are required to establish policies to ensure that recycled oils are acceptable for purchase and are not excluded from state bids. Further, unless it costs more, is unavailable in a reasonable time, or cannot meet performance

standards, the oil that is purchased must contain the greatest percentage of recycled oil available. In addition, the California Integrated Waste Management Board and state officers and employees are encouraged to use recycled oil products.

- Bidders on state contracts must specify the percentage of recycled content in products other than paper, and there is a purchasing preference for recycled over nonrecycled products. The following goals were set regarding the required percentages of recycled products in total state purchases: by 1 January 1991, at least 10 percent; by 1 January 1993, at least 20 percent; and by 1 January 1995, at least 40 percent.
- The California legislature has declared its intent to stimulate market development for recycled products, and it requires that a report be submitted to advise the state on the legislature's effectiveness in achieving this goal. The legislature itself is required to exhibit a purchasing preference for recycled paper products and has specific goals for buying these goods. In addition, state government departments must prepare annual reports for the legislature on the amount of recycled products purchased.
- A Beverage Container Recycling Advisory Committee was established on 1 March 1987 to advise the director of the California Integrated Waste Management Board on all matters concerning the recycling of beverage containers. The California Beverage Container Recycling Fund was established on 29 September 1986 to foster recycling of containers. All dealers who carry recyclable containers must identify, with signage, the nearest certified recycling center, its features, and its hours of operation. There must be at least one certified recycling center within every convenience zone (including reverse vending machines) with required minimum hours of operation. An "antiscavenging law" makes it illegal for anyone other than an authorized recycling agent of the city or county to remove segregated recyclables from collection locations.
- All commercial curbside recycling programs must submit survey forms as required by the state, with information on general operations, actual volumes collected, and revenues received from sales of recyclables.
- Since 1990, each county in California has been required to create a task force at five-year intervals to assist in the development of community source reduction and recycling. All

county and city source reduction and recycling elements "shall place primary emphasis on implementation of all feasible source reduction, recycling, and composting programs while identifying the amount of landfill and transformation capacity that will be needed for solid waste which cannot be reduced at the source, recycled, or composted."

- The state requires that on and after 1 January 1993 "every seller of trash bags of 1.0 mil or greater thickness sold in California shall ensure that at least 10 percent of the material used in these trash bags is recycled postconsumer material." On or after 1 January 1995, the bag thickness requirement was lowered to .75 mil, and the postconsumer content requirement raised to 30 percent. Annual certification of trash bag sellers will be undertaken, as well as audits of sellers as the California Integrated Waste Management Board deems necessary. Problems encountered by sellers after 1 January 1995 with regard to obtaining required postconsumer materials must be certified to the board.
- The California Integrated Waste Management Board is required to review, identify, and report to the Department of General Services any impediments to the procurement of recycled plastic products. A procurement preference has been established for recycled plastic products over those manufactured from virgin resins.
- The Department of Transportation is required to use recycled paving materials as much as possible, and all bid specifications are to be reviewed with regard to the purchase of such materials.
- Since 1991, every newsprint consumer in California has been required to ensure that at least 25 percent of the paper it uses is made from recycled-content newsprint. Goals for the percentages of newsprint made from recycled-content paper are: 30 percent on and after 1 January 1994, 35 percent on and after 1 January 1996, 40 percent on and after 1 January 1998, and 50 percent on and after 1 January 2000. Consumers of newsprint may be prosecuted for fraud should they falsely certify recycled contents.
- A tire recycling program was implemented on 1 July 1991, in order to "promote and develop alternatives to the landfill disposal of used whole tires." The state established grants for research, development, education, and implementation of tire recycling programs. The Department of General Services has been directed to revise its procedures and procure-

ment specifications for the state purchases of any products that could be derived from the recycling of used tires, and it establishes a 5 percent bidding preference. Contracts will be awarded to the bidders whose products include the greatest percentage of recycled tire content, all other things being equal.
- All unused ballots may be recycled after an election. The clerk must make and file a written affidavit stating the number of ballots recycled.

SOURCE: California Education Code: Title 1, Division 1 (General Education Code Provisions), Part 19, Chapter 3 (Miscellaneous), Article 8 (Recycling Paper), §32372, §3273 a–e. California Public Contract Code: Division 2 (General Provisions), Part 2 (Contracting by State Agencies), Chapter 1 (State Contract Act), Article 7 (Contract Requirements), §10233, Chapter 2, Article 7.6 (Recycled Oil Markets), §10406, §10407, Chapter 2.1, Article 2, §10507.5, Chapter 2.5, Article 8.5, §10860, Chapter 4, Article 2, §12162 through §12169, Chapter 4, Article 4, §12205, §12210, §12213, §12225, §12226, Chapter 5, Article 2, §12310, Article 3. California Public Resources Code: Division 3, Chapter 1, Article 9 (Used Oil Recycling Act), §3469, Division 12.1 (California Beverage Container and Litter Reduction Act), Chapter 3, §14530.1, §14531, §14538, §14540, §14542.1, Chapter 4, §14551, §14551.5, Chapter 6, §14570, §14571, Chapter 7, §14580, Division 12.2 (Household Batteries), Division 30 (Waste Management), Part 1, Chapter 2, Part 2, Chapter 2, Article 4, Chapter 3, Article 4, Chapter 6, Article 1, Article 2, Chapter 7, Article 4, Chapter 9, Article 1, §41953, Article 2, §41970, Part 3, Chapter 4, §42202, Chapter 6 (Plastic Recycling Program), Chapter 8 (Recycled Battery Programs), Chapter 10 (Office Paper Recovery Program), Chapter 14 (Paving Materials), Chapter 15 (Newsprint), Chapter 17 (California Tire Recycling Act). California Elections Code: Division 12, Chapter 12, §17136.

Colorado

- As of April 1996, Colorado's recycling rate was 18 percent.
- Colorado requires that "major solid waste disposal sites shall be developed in accordance with sound conservation practices and shall emphasize, where feasible, the recycling of waste materials."
- On public projects, the state provides for a preference of up to 5 percent for bids on finished products that contain no less than 10 percent recycled plastics.

- The state legislature has made a series of declarations related to recycling, one of which stipulates that state government programs should be developed to address the use of recycled products. The legislature has also established a preference for the use of recycled paper in state government contracts.
- A bid preference of 10 percent is allowed for bidders who use recycled paper in the manufacture of commodities or supplies described in any state bid. The legislature expects that paper or paper products bought by the state will be made of recycled paper if the price is competitive and the quality adequate. The following goals for the purchase of recycled paper as a percentage of the total volume of paper and paper products purchased by the state were established: at least 10 percent in fiscal year 1990–1991, 20 percent in 1991–1992, 30 percent in 1992–1993, 40 percent in 1993–1994, and 50 percent in 1994–1995 and for each subsequent fiscal year.
- Another legislative declaration states that "the recycling of plastic materials is a matter of statewide concern and that such recycling should be promoted." This declaration also says that recycling of plastics will decrease the amount disposed of in landfills and will spur economic development in Colorado's recycling industry. The intent of this declaration is "to encourage the development of the recycling industry and the development of markets for recycled plastic materials."
- A third legislative declaration speaks to the importance of proper disposal of solid wastes and recognizes that citizens are concerned about learning how to make maximum use of waste reduction and recycling programs: "Realistic waste reduction goals should be established and state and local solid waste management activities should strive to achieve such goals through source reduction, recycling, composting, and similar waste management strategies."
- Colorado requires that on or after 1 July 1992, all plastic bottles and rigid plastic containers distributed, sold, or offered for sale must be labeled according to the recycling code of the plastics industry. However, "no unit of local government shall require or prohibit the use or sale of specific types of plastic materials or products."
- A pilot program has been established to encourage research and development for new plastics recycling technologies by private industry. The executive director of local affairs is permitted by the state to make grants or loans to private industries for such research and development if the industries are

located in or will expand into Colorado. The executive director is also authorized to accept any grants or loans from private or public sources to apply to such plastics recycling projects.

- The state provides an income tax credit for Colorado residents investing in plastics recycling technologies: "a plastic recycling credit equal to 20 percent of net expenditures to third parties for rent, wages, supplies, consumable tools, equipment, test inventory, and utilities up to ten thousand dollars made by the taxpayer for new plastic recycling technology in Colorado, with a maximum credit of two thousand dollars." A similar tax credit is allowed for domestic and foreign corporations that invest in technologies for recycling plastics. A further tax credit was provided for qualified equipment utilizing postconsumer waste (applicable for income tax years beginning on 1 January 1991 and ending by 1 January 1996).
- Colorado also provides for the adoption of regulations that prohibit discrimination against the transportation of recycled or recyclable materials by rail.

SOURCE: Colorado Revised Statutes: Title 8 (Labor and Industry), Article 19.5 (Bid Preferences—Recycled Plastic Products), §101, Article 19.7 (Bid Preferences—Recycled Paper Products), §101, §102, §103; Title 24 (Government—State), Article 65.1, Part 2, §204.2, Article 103 (Source Selection and Contract Formation), Part 2 (Methods of Source Selection), §207; Title 25 (Health; Environmental Control), Article 17 (Recycling of Plastics), §101–106; Title 30 (Government—County; County Powers and Functions, General), Article 20, Part 1 (Solid Wastes Disposal Sites and Facilities), §100.5, §101, §102.5; Title 39 (Taxation; Specific Taxes; Income Tax), Article 22 (Income Tax), Part 1, §114.5, Part 3 (Corporations), §309, Part 5 (Special Rules), §515(f); Title 40 (Utilities; Public Utilities; General and Administrative), Article 3, §113.

Connecticut

- As of April 1996, Connecticut's recycling rate was 23 percent.
- A plan to increase state procurement of goods that contain recycled materials was first required on 1 October 1989 and is expected to be updated periodically. The plan was also expected to include specifications for postconsumer and secondary waste content in recycled products.
- Further plans were mandated in 1989 to reduce the use of disposable and single-use products, as well as to separate and

collect items deemed suitable for recycling. Annual progress reports from state agencies on the implementation of such plans are required.

- The state is also directed by the legislature to use two-sided copies whenever possible to reduce paper waste at its source. White paper recycling programs were instituted within the legislature itself in 1989, and all other state agencies were required to follow suit by 1991.
- All state bids must be accompanied by statements that assess the recyclability of materials. Furthermore, the commissioner of transportation must ensure that materials specifications for state transportation projects encourage the use of recycled materials.
- Since 1989, all owners/operators of resource recovery facilities and solid waste disposal areas have been required to submit quarterly reports detailing the amounts of solid waste received. Recycling facilities are also required to submit similar reports for the following recyclable items: cardboard; leaves; glass, plastic, and metal food and beverage containers; newspapers; storage batteries; and waste oil.
- The state also mandates that plans be submitted and approved by the commissioner for environmental protection for the disposal or recycling of ash residue from any incinerators or resource recovery facilities.
- Facilities that compost leaves are regulated by the state, but leaf composting and recycling facilities are exempt from the requirement to obtain a solid waste facility permit.
- All municipal authorities must provide for the safe and sanitary disposal of nonhazardous solid waste (disposal of hazardous waste is the responsibility of the generator). Also, municipal authorities are allowed to inspect any recycling facilities that handle municipal wastes. As of 1 July 1993, each municipality is expected to provide for the recycling of nickel-cadmium batteries contained in consumer products. Subsequently, on 1 May 1996, another law went into effect making it mandatory to recycle nickel-cadmium batteries.
- State law requires the establishment of municipal solid waste recycling programs. A trust fund has been created to provide municipalities with grants to assist their recycling program development. Any revenue collected through municipal solid waste programs must be deposited into this fund. A goal of recycling not less than 25 percent of the solid waste generated in the state after 1 January 1991 was set. Since that date,

state residents have been required to source-separate recyclable items in their solid waste at home.

- The state is also required to: (1) coordinate litter control and recycling programs in state and local agencies; (2) develop public education programs about litter and recycling; (3) encourage, organize, and coordinate voluntary antilitter and recycling campaigns within communities; (4) investigate the availability of funding for such programs; and (5) study current research and developments in litter control and recycling.
- There is also a state plan for the promotion and support of industries using recycled materials. Progress reports are required at six-month intervals.
- Official symbols and the procedures to use them have been established to indicate that packaging is recyclable or made from recycled materials.
- Goals for the percentages of recycled fiber in the newsprint used by all publishers and other printers were established as follows: 11 percent or more for the year ending 31 December 1992, 16 percent or more for the year ending 31 December 1993, 20 percent or more for the year ending 31 December 1994, 23 percent or more for the year ending 31 December 1995, 31 percent or more for the year ending 31 December 1996, 40 percent or more for the year ending 31 December 1997, 45 percent or more for the year ending 31 December 1998, and 50 percent or more for the year ending 31 December 1999 and thereafter.
- Goals were also established for telephone directory publishers regarding the use of recycled fiber in directory stock: 10 percent or more for the year ending 31 December 1995, 15 percent or more for the year ending 31 December 1996, 20 percent or more for the year ending 31 December 1997, 25 percent or more for the year ending 31 December 1998, 30 percent or more for the year ending 31 December 1999, 35 percent or more for the year ending 31 December 2000, and 40 percent or more thereafter. All directory publishers were required to file a plan on or before 1 January 1991 that provided that at least 10 percent of their directories would be collected and recycled. This percentage must be increased by 5 percent of the total annually until 50 percent or more of all directories are retrieved and recycled.
- In 1994, Connecticut was one of the first states in the nation to revise its specifications for the purchase of printing and writing papers to incorporate the more stringent standards

of the Environmental Protection Agency's paper procurement guidelines, as specified in Executive Order 12873, including the requirements for minimum recycled content.

- Also in 1994, the state legislature made it possible for municipalities to impose a fine up to $1,000 for the violation of solid waste disposal laws.
- In 1995, a bill banning the disposal of grass clippings was approved, but the effective date was postponed until October 1997 to allow time for the Department of Environmental Protection to develop regulations for grass composting, as well as to establish pilot projects in communities throughout the state.

SOURCE: Connecticut General Statutes: Title 4A (Administrative Services), Chapter 58 (Purchases and Printing), §67a; Title 4B (State Real Property), Chapter 59 (Department of Public Works and State Real Property), Part II, §15; Title 13A (Highways and Bridges), Chapter 238 (Highway Construction and Maintenance), Part V, §981; Title 22A (Environmental Protection), Chapter 446D (Solid Waste Management), §208e, §208g, §208i, §220, §241, §241a, §241b, §241g, §241h, §249, §255c, §256a, §256n, §256p, §256t, §256z, §256ee; Title 32 (Commerce and Economic Development), Chapter 578 (Department of Economic Development), §1e. Public Acts No. 94-153, 94-200. State Senate Bill 1052 (1995).

Delaware

- As of April 1996, Delaware's recycling rate was 33 percent.
- Delaware has declared that its citizens have a right to a clean and wholesome environment, that solid waste disposal practices in the past have not been conducive to this end, and that locally organized recycling programs have demonstrated that solid wastes generated within the state do contain recoverable resources.
- The General Assembly has established that: (1) "maximum resources recovery from solid waste and maximum recycling and reuse of such resources in order to protect, preserve and enhance the environment of the State shall be considered environmental goals of the State"; (2) "solid waste disposal and resource recovery facilities and projects are to be implemented either by the State or under state auspices, in furtherance of these goals"; (3) "appropriate governmental structure, processes and support are to be provided so that

effective state systems and facilities for solid waste management and large-scale resources recovery may be developed, financed, planned, designed, constructed and operated for the benefit of the people, municipalities and counties of the State"; (4) private industry is to be utilized to the fullest extent to assist in goals of resource recovery, recycling, and reuse; and (5) "long-term contracts are authorized between the State and private persons or businesses to further resource recovery."

- To meet these goals, the Delaware Solid Waste Authority was created through the Waste Minimization and Recycling Acts of 1990. The authority is empowered to adopt rules and regulations governing the composition, quantity, quality, and transportation of source-separated recyclables to recycling centers and to develop a statewide recycling and waste reduction plan. The authority is also expected to establish recycling centers in each county, to provide for public education about source-separated recycling and resources recovery, to promote markets for recyclable materials, and to report annually on its activities.
- The General Assembly has also determined that beverage containers are "a major source of nondegradable litter" in the state and that litter control is a financial burden. With the intention of creating incentives to reuse or recycle beverage containers, the state has mandated the establishment of redemption centers, which have also been empowered to accept materials other than beverage containers for reuse or recycling.
- Delaware prohibits metal beverage containers with any detachable part; container holders, such as six-pack rings, that are not classified by the state as biodegradable, photodegradable, or recyclable; and glass beverage containers that are not recyclable or refillable.
- Recyclable materials are defined by Delaware to include newsprint, computer paper, white paper, corrugated and other cardboards, plastics, ferrous metals, nonferrous metals, white goods, organic yard waste, used motor oil, asphalt, batteries, and household paint, solvent, pesticide, and insecticide containers.

SOURCE: Delaware Code Annotated: Title 7 (Conservation), Part VII (Natural Resources), Chapter 60 (Environmental Control), Subchapter III (Beverage Containers), §6051, §6052, §6056, §6059,

Chapter 64 (Delaware Solid Waste Authority), Subchapter I, §6401, §6402, §6403, §6404, §6406, §6407, Subchapter II (Recycling and Waste Reduction), §6450–6459, Chapter 78 (Waste Minimization/ Pollution Prevention Act), §7802–7804. Title 21 (Motor Vehicles) delineates numerous laws that apply specifically to automotive recyclers and the automotive recycling industry.

District of Columbia

- As of April 1996, the District's recycling rate was 15 percent.
- The mayor of the District of Columbia was directed by statute to submit to the District's council, within one year of 25 July 1987, a comprehensive plan for a multimaterial recycling system for the purpose of "recovering energy and other resources from discarded materials and solid waste and for distributing reusable organic compounds for public use."
- A District Office of Recycling was subsequently established. Additionally, a priority for recycling was defined, and the District stated that no incineration facilities would be constructed until all the provisions of the D.C. Solid Waste Management and Multi-Material Recycling Act of 1988 (D.C. Law 7-226) were implemented or a 25 percent reduction in the solid waste stream was achieved through recycling, whichever came first.
- This act, which took effect on 1 October 1989, required every household to separate newspaper, yard waste, metals, and glass for recycling. A requirement regarding the collection of some plastics was added on 8 June 1991. The act also states that commercial businesses must recycle newspapers, metals, glass, and office paper.
- District goals included the recycling of at least 35 percent of the total solid waste stream by 1 October 1992 and the recycling of at least 45 percent of the total solid waste stream by 1 October 1994.
- The mayor was also required to establish at least one multimaterial buy-back center in the District on or before 1 October 1989, as well as at least one intermediate processing facility to receive recyclable materials.
- The District requires that newspapers, metal food and beverage cans, glass food and beverage containers, and yard wastes are to be source separated by residents for collection by commercial services (although people may also sell/donate their recyclables to drop-off or buy-back centers).

- The use of compost by the District's land maintenance programs is encouraged.
- In 1995, an amendment was passed imposing a recycling surcharge on "all persons who dispose of solid waste through the solid waste disposal system of the District to offset the cost of developing new and additional methods of solid waste management." Monies from this fee were to be used for funding recycling activities in the District, "no more than 25 percent of which shall go to fund the recycling educational and promotional activities of the Environmental Planning Commission."
- There is a Tire Recycling Fee of $2 for each new tire sold in the District, and the monies generated are to be pooled and used with the monies from the recycling surcharge.
- Minimum recycled-content percentage requirements are also defined for paper or paper products: Beginning 1 January 1994, all persons selling or distributing significant quantities of paper or paper products in the District were required to ensure that these goods contain at least the minimum percentage of recycled content designated by the Environmental Protection Agency in the Resource Conservation and Recovery Act of 1976, as set forth in the Code of Federal Regulations (40 CFR 250.21).
- Newsprint must be made of at least 40 percent postconsumer recovered materials, and high-grade bleached printing and writing papers must contain at least 50 percent wastepaper.
- All corporations that file District annual reports and are sellers or distributors of paper or paper products must include information on the minimum recycled content of papers sold or distributed per District requirements. As of 1 January 1994, a minimum recycled-content surcharge is applied to any person who fails to comply with the District's recycled paper requirements. All revenues received from this surcharge will be applied to District recycling activities.

SOURCE: District of Columbia Code: Part I (Government of District), Title 6, Chapter 32 (Multi-Material Recycling Systems), §3202, Chapter 34 (Solid Waste Management and Multi-Material Recycling), §3401, §3405–3410, §3415, §3419, §3420, §3422, §3423. Paper and Paper Products Recycling Incentive Amendment Act of 1990 (D.C. Law 8-208). Recycling Fee and Illegal Dumping Amendment Act of 1995 (D.C. Law 11-12).

Florida

- As of April 1996, Florida's recycling rate was 40 percent.
- A Florida Seed Capital Fund has been established "to provide equity financing for the research and development activities of new and existing high technology small business in the state." "High technology" businesses include those that provide products or services in the area of recyclable materials.
- When the State Comprehensive Plan was adopted in 1988, one of its goals was the initiation of programs to "develop or expand recyclable material markets, especially those involving plastics, metals, paper, and glass." The plan also encouraged "the research, development, and implementation of recycling, resource recovery, energy recovery, and other methods of using garbage, trash, sewage, slime, sludge, hazardous waste, and other waste."
- Machinery and equipment used for processing recyclable materials are exempt from sales tax, and no sales tax is added to recycled or waste oils and solid waste materials if they are used as fuels.
- The state has permitted the Department of Education to contract recycling firms to pick up, process, and recycle obsolete or unusable materials from the schools.
- A program has been established to award contracts and grants to independent, nonprofit colleges and universities for research activities related to methods and processes for recycling solid (and hazardous) waste. Florida also encourages collaboration between private industry and state universities on high-technology research projects, including projects to resolve problems associated with the design and implementation of programs to "recycle materials such as plastics, rubber, metal, glass, paper, and other components of the solid waste stream."
- State agencies are directed to purchase recycled paper (when economical), and bidders on printing contracts must certify the percentages of recycled content in their papers. A bidding preference is provided for recycled papers.
- The Division of Purchasing is prohibited from discriminating against products and materials with recycled content.
- The state also encourages the Department of Transportation to increase the use of recovered materials in its construction programs.

- The legislature has stated that it will "promote the recovery of energy from wastes," including the recycling of manufactured products.
- A legislative declaration states that "maximum resource recovery from solid waste and maximum recycling and reuse of such resources must be considered goals of the state." The legislature also has been directed to promote the reduction, recycling, reuse, or other alternative treatment of solid waste in lieu of disposal. Moreover, state agencies are expected to develop, promote, and implement public education programs about solid waste issues and recycling, in addition to developing and implementing actual recycling programs.
- The powers and duties of the state's Department of Public Health are delineated by statute, and these include promoting "the planning and application of recycling and resource recovery systems" for the betterment of the environment, as well as for energy recovery. The department is also expected to "assist in and encourage, as much as possible, the development within the state of industries and commercial enterprises which are based upon resource recovery, recycling, and reuse of solid waste." The department's other duties include maintaining a state directory of recycling businesses, managing grants for recycling programs, and increasing public education and awareness about recycling, volume reduction, and proper methods of solid waste disposal.
- Recycling is an integral part of the state's solid waste management effort, and each county was directed to initiate a materials recycling program by 1 July 1989. These programs required the counties to separate construction and demolition debris; separate from the waste stream and offer for recycling the majority of newspapers, aluminum cans, and glass and plastic bottles; encourage local governments to recycle plastics, metals, and all grades of paper; recycle and/or compost yard trash and mechanically treated solid waste; and ensure that the county's total municipal solid waste would be reduced by at least 30 percent by the end of 1994. No more than half of this goal could be met through recycling yard trash, white goods, construction/demolition debris, and tires.
- As of 1 October 1989 and on an annual basis thereafter, each county was required to submit a report to the Department of Public Health detailing its solid waste management program and recycling activities.

- The procurement of products or materials with recycled content is also addressed, with the state requiring that such products be used whenever possible. Preconsumer waste is specifically excluded from the state's definition of "recycled content."
- Penalties and prohibitions with respect to solid wastes and recycling are provided. For example, the following disposal bans have been established: After 1 October 1988, used oil could no longer be disposed of in landfills; after 1 January 1989, lead-acid batteries could not be disposed of in landfills, and all persons who sell such batteries must also collect trade-ins for recycling; after 1 January 1990, white goods could not be accepted at landfills; and after 1 January 1992, yard trash could not be disposed of in sanitary landfills.
- By 1 September 1989, all state agencies were required to establish recycling programs for the recovery of, at a minimum, aluminum, high-grade office paper, and corrugated cardboard.
- Florida encourages the development of recycled materials markets, and, since 1 September 1989, it has been mandatory that a report assessing the recycling industry and its markets be submitted on an annual basis.
- A grant program has been developed for municipalities and counties to implement solid waste management recycling and education programs. The state has also directed the school board of each district to provide instructional programs on recycling to all students at both the elementary and secondary levels. Further, the Department of Education has been assigned the task of developing curriculum materials and resource guides for recycling awareness programs for kindergarten through twelfth-grade classes.
- Another statute directed the capitol, the house, and the senate office buildings to institute recycling programs for wastepaper and aluminum cans as of 1 January 1989.
- A products waste disposal fee of 10 cents per ton on all newsprint consumed by publishers within the state took effect after 1 January 1989, and a 10-cent credit per ton is provided for the use of recycled newsprint. If it is determined on 1 October 1992 that newsprint sold within the state is being recycled at a rate of 50 percent or more, then the disposal fee will be rescinded. However, if the recycling rate is less than 50 percent, the disposal fee and corresponding credits will be increased to 50 cents per ton on that same date. Additionally,

if the rate is less than 50 percent on 1 October 1992, "any producer or publisher using newsprint in publications shall accept from a person for recycling purposes reasonably clean newsprint previously produced, published, or offered for sale by that producer or publisher." A credit of 25 cents per ton of newsprint used is allowed if a publisher can prove that such paper was recycled through its facility.

- If a sustained recycling rate of 50 percent has not been reached by 1 October 1992 for containers made of glass, plastic, plastic-coated paper, aluminum, or other metals, then an advance disposal fee of one cent per container shall be charged by retailers. The proceeds from this advance disposal fee are to be deposited into the Container Recycling Trust Fund. Containers for which such a fee has been charged may be returned to recycling centers for a refund of that fee in addition to any payment for market value. If a recycling rate of 50 percent for such containers has not been reached by 1 October 1995, then the advance disposal fee will be increased to two cents per container. Such fees will not apply to any containers that are being recycled at a rate of 50 percent or more, and refillable containers are excluded as well.
- Although the legislation states that "each consumer shall deposit with the dealer the refund value of each container purchased," containers sold for consumption on the premises of the dealer are exempt from this deposit. Dealers must accept from the public any empty, unbroken, and reasonably clean container of the type, size, and brand they have sold within the past 60 working days. Distributors will not be required to pay deposits to manufacturers on nonrefillable containers. The legislation also provides for the establishment of redemption centers.
- Florida has examined the issue of hazardous household wastes and has decided to require the establishment of programs to handle such materials. Thus far, however, only used oil has been addressed. According to state law, used oil may only be disposed of in a manner consistent with recycling or beneficial reuse (this specifically excludes its use for road oiling, dust control, weed abatement, or other similar applications that have the potential to release used oil into the environment). The state also requires the development and implementation of a public education program about collecting and recycling used oil, and it encourages the voluntary establishment of facilities to collect used oil for recycling. A 5

percent price preference for the purchase of recycled oils by state and local governments has been created to encourage the industry.

SOURCE: West's Florida Statutes Annotated: Title XI (County Organization and Intergovernmental Relations), Chapter 159 (Bond Financing), Part III, §159.445; Title XIII (Planning and Development), Chapter 187 (State Comprehensive Plan), §187.201; Title XIV (Taxation and Finance), Chapter 212 (Tax on Sales, Use, and Other Transactions), Part I, §212.08; Title XVI (Education), Chapter 233 (Courses of Study and Instructional Aids), §233.37; Chapter 240 (Postsecondary Education), Part V, §240.5325, §240.539; Title XIX (Public Business), Chapter 283 (Public Printing), Part I (Executive Agency Printing), §283.32, Chapter 287 (Procurement of Personal Property and Services), Part I, §287.045; Title XXVI (Public Transportation), Chapter 336 (County Road System), §336.044; Title XXVIII (Natural Resources; Conservation, Reclamation, and Use), Chapter 377 (Energy Resources), Part II, §377.703; Title XXIX (Public Health), Chapter 403 (Environmental Control), Part IV (Resource Recovery and Management), §403.702–403.7065, §403.708, §403.7095, §403.714, §403.7145, §403.717, §403.7195, §403.7197, §403.7198, §403.7265, §403.751, §403.753; Title XXXIII (Regulation of Trade, Commerce, Investments, and Solicitation), Chapter 538 (Secondhand Dealers and Secondary Metals Recyclers), all parts and sections within this chapter regulate secondary metals recyclers.

Georgia

- As of the end of 1995, Georgia's recycling rate was calculated to be 32.6 percent.
- In 1990, Georgia announced a policy to institute and maintain a comprehensive state solid waste management plan. This policy directed state agencies to make plans to educate and encourage solid waste handlers and generators to utilize source reduction, reuse, composting, recycling, and other alternative waste management methods to cut the amount of solid waste.
- The state has also declared its intention to promote markets for and engage in the purchase of products made from recovered materials and recyclable goods. A Recycling Market Development Council was established on 1 July 1990. At the same time, a review of purchases and purchasing specifica-

tions, practices, and procedures with respect to recycled materials was begun.

- In 1993, a reorganization of state agencies occurred and four agencies were charged with specific duties with regard to solid waste management: the Environmental Protection Division and the Pollution Prevention Assistance Division of the Georgia Department of Natural Resources, the Georgia Department of Community Affairs, and the Georgia Environmental Facilities Authority.
- A statewide goal was set to reduce the amount of municipal solid waste disposed of in 1992 by 25 percent as of 1 July 1996.
- Georgia has directed each city and county to develop or be included in a comprehensive solid waste management plan. After 1 July 1992, no permits for solid waste facilities will be issued unless the applicant's host jurisdiction can, among other things, demonstrate an active involvement and strategy for meeting the statewide goal for the reduction of solid waste disposal by 1 July 1996. State authorities have also been empowered to establish recycling/resource recovery programs to help the public achieve this end. In addition, there is a statute encouraging the recovery and utilization of resources contained in solid waste and sewage sludge.
- In 1994, the state required that as of 1 September 1996, each local government shall have established yard trimmings disposal prohibitions and regulations.
- After 1 January 1991, lead-acid batteries may not be disposed of in municipal solid waste landfills but must be delivered to some sort of materials recovery program. In addition, any person selling lead-acid batteries must also accept them for recycling and post signage with specified language about the law as it pertains to lead-acid batteries.
- All cities, counties, and solid waste management authorities were authorized to impose restrictions on the disposal of tires as of 1 July 1990. Such restrictions may include a ban on tire disposal and a requirement that tires "be recycled, shredded, chopped, or otherwise processed in an environmentally sound manner prior to disposal."
- Since 1 January 1991, the state has required that rigid plastic containers or bottles manufactured or sold in Georgia must be labeled (per the specifications of the Society of the Plastics Industry).

- Georgia grants an 8 percent price incentive for the purchase of paper products made from recovered fiber over similar products made from 100 percent virgin fibers. However, the state limits the total amount expended on such products to $1 million per year.
- In the Georgia Business Expansion & Support Act of 1994, investment tax credits were begun for companies investing in recycling, pollution control, and defense conversion activities.
- In March 1997, a bottle deposit bill will be brought before the state legislature. When the legislative session ended in April, the bill had not yet been voted on. The bill will be reintroduced when the legislature reconvenes in January 1998.

SOURCE: Code of Georgia: Title 12 (Conservation and Natural Resources), Chapter 8 (Waste Management), Article 2 (Solid Waste Management), Part 1 (General Provisions), §12-8-21, §12-8-28, §12-8-31.1, §12-8-33, §12-8-34, §12-8-35, §12-8-40.1, §12-8-40.2, Part 2 (Regional Solid Waste Management Authorities), §12-8-51; Title 36 (Local Government), Chapter 63 (Resource Recovery Development Authorities), §36-63-2; Title 50 (State Government), Chapter 5 (Department of Administrative Services), Article 3 (State Purchasing), Part 1 (General Authority, Duties, and Procedure), §50-5-60.1, §50-5-60.2. H.B. 1527, 1994. Georgia Department of Community Affairs.

Hawaii

- As of April 1996, Hawaii's recycling rate was 20 percent.
- Hawaii has established pricing preferences for bidders offering goods containing recycled materials. Rules establishing percentages and preference guidelines were to be established by the comptroller by 1 January 1992.
- One of the state's legislative objectives is to "promote re-use and recycling to reduce solid and liquid wastes and employ a conservation ethic."
- Another statute establishes that the director of the Department of Public Health is responsible for the study of available research in the field of recycling, the study of methods for implementing such research, and the development of public educational programs.
- Hawaii prohibits the land disposal of lead-acid batteries and requires that they be recycled. Additionally, all sellers of such batteries must post the relevant law and accept used batteries (for recycling) from customers buying new ones.

- The state also prohibits the disposal of any new, used, or recycled oil either on the ground or in the water.
- The state environmental policy guidelines call for the promotion of "the optimal use of solid wastes through programs of waste prevention, energy resource recovery, and recycling so that all our wastes become utilized."

SOURCE: Hawaii Revised Statutes Annotated: Title 9 (Public Property, Purchasing, and Contracting), Chapter 103 (Expenditure of Public Money and Public Contracts), Part II, §103-24.5; Title 13 (Planning and Economic Development), Chapter 226 (Hawaii State Planning Act), Part I, §226-15; Title 19 (Health), Chapter 339 (Litter Control), Part I, §339-1, §339-3, Chapter 340A (Solid Waste), §340A-1, Chapter 342I (Lead-Acid Battery Recycling), §342I-1, §342I-2, Chapter 342N (Used Oil Transport, Recycling, and Disposal), §342N-3, Chapter 344 (State Environmental Policy), §344-4.

Idaho

- Idaho's solid waste and recycling laws are presently limited to two disposal bans. One requires that all used tires be recycled and that any person who sells tires must post this law and accept used tires for recycling from anyone who purchases new or retreaded tires.
- The second ban prohibits the incineration or disposal of lead-acid batteries in landfills. Such batteries must be recycled, and all sellers of lead-acid batteries must post the relevant law. Also, all lead-acid batteries sold after 1 July 1992 must bear a universally accepted recycling symbol.

SOURCE: Idaho Code: Title 39 (Health and Safety), Chapter 65 (Waste Tire Disposal) §39-6502, Chapter 70 (Sale and Disposal of Batteries), §39-7002, §39-7003.

Illinois

- As of April 1996, Illinois' recycling rate was 27 percent.
- Illinois permits local governments to contract with other local governments, private corporations, or nonprofit organizations for the purposes of garbage disposal or recycling.
- All counties with populations over 100,000 and the city of Chicago were required to develop comprehensive, 20-year waste management plans that focused on recycling,

composting, waste reduction, and other landfill alternatives by 1 March 1991. County program goals included the recycling of 15 percent of municipal solid waste by the third year in operation and 25 percent of such waste by the fifth year (subject to viable markets).

- A total of 25 assistance grants are available to municipalities or combinations thereof with total populations of 20,000 or more for the purpose of implementing pilot recycling programs. At a minimum, such programs must include curbside collection for at least three source-separated materials (the three materials could be chosen from glass, aluminum, steel and bimetallic cans, newsprint, corrugated cardboard, used motor oil, and plastics); a drop-off or buy-back center for at least glass, aluminum cans, and newsprint; provisions for recycling collected materials; and provisions for a public education program and enforced compliance.
- Effective 1 July 1990, all yard waste (grass, leaves, and brush) was banned from landfills, and counties were required to begin composting programs.
- Lead-acid batteries were banned from landfills as of 1 September 1990; tires were similarly banned as of 1 July 1994.
- Effective 1 July 1994, white goods were banned from landfills, unless the "white good components have been removed," i.e., the devices that might contain PCBs, CFCs, or other toxic substances.
- A Vehicle Recycling Board was created by the state to oversee vehicle recycling. All monies collected in the Vehicle Recycling Fund are transferred to the Common School Fund.
- Illinois has a Waste Oil Recovery Act that promotes the recycling of used oil, encourages the purchase of recycled oils, and establishes educational programs designed to inform the public about recycling oil in order to conserve natural resources. As of 1 July 1996, it became illegal for anyone to knowingly mix liquid used oil with any other municipal waste intended for collection and disposal at a landfill.
- The state also has a Recycled Newsprint Use Act requiring that on and after 1 January 1991, consumers of newsprint must meet the following goals for annual recycled fiber usage: 22 percent beginning 1 January 1991, 25 percent beginning 1 January 1992, and 28 percent beginning 1 January 1993. The act also states that if these goals are not met by 1993, every consumer of newsprint in Illinois during 1994 must ensure that its recycled fiber usage is at least 28 percent unless (1)

the required newsprint cannot be found in sufficient quality or quantity or (2) contracts for procurement of newsprint were made before 1 January 1991. Criminal penalties are delineated for violations of this act. At the end of 1994, all goals were met and/or exceeded at each level of this bill. However, a discussion began when the Act reached the end of its authority (1994) as to whether the intent of the bill was to maintain these standards for the future. The state argued that this, indeed, was the intent of the bill; state and national newspaper organizations, on the other hand, have argued that the bill's authority did not, and does not, extend beyond 1994. This discussion is still in progress, although as of 1997 the newspaper interests have prevailed.

- The Illinois Solid Waste Management Act includes a recycling program stipulating that all state agencies that maintain public lands must give preference to using composted materials and that waste reduction and recycling programs for office paper, corrugated cardboard, newsprint, and mixed paper must be instituted in all state buildings. The goals for these programs were a 25 percent reduction in such waste by 31 December 1995 and a 50 percent reduction by 31 December 2000. In addition, procurement procedures and specifications were to be reviewed and modified, where feasible, to encourage the maximum purchase of products made from recycled materials, especially those with the highest content of postconsumer waste. Goals for the percentage of the total amount of paper and paper products purchased by the state that contain recycled materials were: at least 10 percent by 30 June 1989, at least 25 percent by 30 June 1992, and at least 40 percent by 30 June 1996. As of 1 July 2000, this percentage will increase to 50 percent. Furthermore, all paper purchased for use by state agencies must be recyclable whenever possible, and such agencies are to review their paper requirements to allow the use of such paper whenever possible.
- On 29 June 1995, Gov. Jim Edgar signed Senate Bill 336, which ordered that the functions of several natural resource agencies be combined for the purpose of greater efficiency. At the same time, the Solid Waste Management Act formally established the Environmental Protection Agency's waste management hierarchy within the state.
- The Solid Waste Management Act also required state-supported colleges and universities to develop comprehensive waste reduction plans by 1 January 1995, and these plans

were required to include recycling and waste reduction measures that would achieve at least a 40 percent reduction in the amount of solid waste generated by these institutions by 1 January 2000.

- Additionally, as of 31 December 2000, all state agencies are expected to have reduced their waste generation by 50 percent.

SOURCE: Smith-Hurd Illinois Annotated Statutes: Chapter 34 (Counties Code), Article 5, Division 5-1, §5-1048, Chapter 85 (Local Government—Solid Waste Planning and Recycling Act), §5956, §5958, Chapter 95 1/2 (Motor Vehicles), Chapter 4, Article III (Vehicle Recycling Board), §4-302, §239.41, Chapter 96 1/2 (Natural Resources—Waste Oil Recovery Act), §7702.3, §7707, §7708, §7709, Chapter 96 1/2 (Recycled Newsprint Use Act), §9752.25, §9752.30, §9752.35, §9752.45, §9753, §9763, Chapter 111 1/2 (Public Health and Safety—Environmental Protection Act); Title I, §1003.30, Chapter 111 1/2 (Solid Waste Management Act), §7053, §7056a, Chapter 127 (State Government—Illinois Purchasing Act), §132.6-4. Illinois Solid Waste Management Act (415 ILCS 20/1 et seq.); Natural Resources Act (20 ILCS 1105/1 et seq.); Public Acts 85-1430, 86-723, 86-452, 87-858, 87-1213.

Indiana

- As of April 1996, Indiana's recycling rate was 19 percent.
- Indiana has a Recycling Promotion and Assistance Fund designed "to promote and assist recycling throughout Indiana by focusing economic development efforts on businesses and projects involving recycling."
- The Department of the Environment is directed by statute to educate students, consumers, and business about the benefits of solid waste recycling and source reduction and to establish a recycled paper task force to develop voluntary guidelines concerning newsprint and other paper products.
- As of 1 January 1992, each plastic bottle and rigid plastic container must be imprinted with the recycling code that identifies its resin (per the Society of the Plastics Industry specifications).
- State colleges and universities are directed to collect recyclable paper and procure recycled paper products when economically feasible.

- In offering economic development assistance, the state Department of Commerce is directed to give priority to businesses and industries whose primary activities are aimed at converting recyclable materials into useful products.
- State agencies are expected to collect and recycle paper and paper products, as well as procure the same, when economically feasible. The state defines recycled materials as those containing at least 10 percent postconsumer/postmanufacture waste.
- Indiana's 20-Year Solid Waste Management Plan was designed to reduce landfilled solid waste 35 percent by 1995 and 50 percent by 2000.

SOURCE: West's Annotated Indiana Code: Title 4 (State Offices and Administration), Article 13, Chapter 4.1 (Printing for State Agencies), §4-13-4.1-5 , Article 13.4 (State Procurement), Chapter 4, §4-13.4-4-7, Article 23, Chapter 5.5 (Indiana Energy Development Board), §4-23-5.5-14; Title 13 (Environment), Article 7 (Environmental Management), Chapter 3, §13-7-3-6.1, §13-7-3-15, Chapter 22, §13-7-22-1, Article 9.5 (Solid Waste Management), Chapter 1, §13-9.5-1-24; Title 20 (Education), Article 12, Chapter 67 (Recycled Paper Products), §20-12-67-2, §20-12-67-3; Title 24 (Trade Regulations; Consumer Sales and Credit), Article 5, Chapter 17 (Environmental Marketing Claims).

Iowa

- As of April 1996, Iowa's recycling rate was 28 percent.
- Iowa has directed its procurement agencies to purchase recycled paper products whenever possible if the price is "reasonably competitive" and the product is of "the quality intended."
- Soybean-based inks and starch-based plastics (which use two of the state's agricultural products—soybeans and corn) are promoted.
- State agencies were required to establish wastepaper recycling programs by 1 January 1990.
- A program was established to increase the recycling of food service and packaging items by 25 percent as of 1 January 1992 and by 50 percent as of 1 January 1993. If these goals were not met, the manufacturing, sale, and use of polystyrene packing products or food service items would be prohibited

as of 1 January 1994. Iowa has directed its department of general services to comply with the recycling goal and schedule, as well as the ultimate termination of the use of polystyrene products for storing, packaging, or serving food for immediate consumption. This law was repealed in 1995.

- A statewide waste reduction and recycling network has been established to promote the state's waste management policy, which encourages the reduction of waste volume, the development of recycling markets, and the education of the public to that end. A recycling demonstration project for polystyrene food packaging was to be implemented by 1 July 1991, and a report on the results was to be submitted to the General Assembly by 1 January 1992. The report was completed and submitted on time.
- As of 1 July 1990, land disposal of lead-acid batteries was prohibited. Battery retailers must post this law and accept old batteries for recycling when new ones are purchased.
- Iowa has banned the disposal and required the recycling of the following materials: beverage containers (1990), lead-acid batteries (1990), used oil (1990), tires (1991), yard waste (1991), and nondegradable grocery bags (July 1992).
- A Waste Volume Reduction and Recycling Fund has been established to award grants for the implementation of a pollution hotline and for projects dealing with waste reduction, recycling, and education about such issues.

SOURCE: Iowa Code Annotated: Title II, Chapter 18, Division 1, §18.18, §18.20, §18.21; Title XVII (Natural Resource Regulation), Chapter 455D (Waste Volume Reduction and Recycling), §455D.5, §455D.10, §455D.15, §455D.16.

Kansas

- As of April 1996, Kansas' recycling rate was 8 percent.
- Kansas allows a bidding preference for products containing the highest percentage of recycled materials, but the bid must be competitive with others. Specific statutes establish exact percentages of total dollar amounts to be spent on recycled paper. In addition, state agencies are not permitted to discriminate against products containing recycled materials.
- In 1990, a statute established a statewide coordinator position to deal with waste reduction, recycling, and market development. That position was abolished as of 1 July 1995.

- The Kansas Commission on Waste Reduction, Recycling, and Market Development was established on 1 July 1990 to evaluate and recommend specific actions that the governor and legislature may take to reduce the volume of generated solid waste, expand markets for recyclable materials, encourage local recycling and waste reduction programs, promote recycling, and create opportunities for recycling enterprises. After submitting a report on its findings, the commission was abolished on 1 July 1992.
- Kansas prohibits the land disposal of used tires. As of 1 July 1991, programs were established to enforce laws pertaining to tire disposal and to encourage the recycling of tires. Grant monies are available to fund research and development on recycling and reusing waste tires, and in 1996 four new tire grant programs were created by the Kansas Legislature.

SOURCE: Kansas Statutes Annotated: Chapter 74, Article 50, §74-5087, §74-5088, Chapter 75, Article 37, §75-3740, §75-3740b, §75-3740c, Chapter 65 (Public Health), Article 34 (Solid and Hazardous Waste), §65-3406, §65-3418, §65-3424, §65-3424f, §65-3431. Recycling of vehicles and vehicle parts is addressed in Chapter 8, Article 1, §8-1,136, §8-1,137, and Article 24, §8-2401, §8-2401a, §8-2406. S.B. 399, 1996.

Kentucky

- As of April 1996, Kentucky's recycling rate was 15 percent.
- Kentucky prohibits discrimination against the purchase of recycled materials and has required all state agencies to establish minimum recycled-content purchasing regulations since 1 September 1991.
- Any bidder competing for a state contract and any projects financed by state bonds must use products containing recycled materials, per minimum content regulations, wherever possible.
- Products with minimum recycled content must be available through the state's central stores. Any state agency that purchases these goods will receive a 50 percent reduction in any administrative fees normally charged by the central purchasing agency.
- Kentucky offers a tax credit equal to 50 percent of the installed cost for recycling or composting equipment.

- A Kentucky Recycling Brokerage Authority was established on 1 July 1991 to further local governments' efforts to develop reliable markets for their recyclables.
- Twenty percent of the funds available to the state bond program must be used for projects that create or expand markets for materials recovered or diverted from the solid waste stream.
- The state prohibits the disposal of used oil and requires that it be recycled or reused.

SOURCE: Kentucky Revised Statutes Annotated: Title VI, Chapter 45A (Recycled Material Content Products), §45A.515, §45A.520, §45A.525, §45A.530, §45A.540; Title XI (Revenue and Taxation), Chapter 141, §141.390; Title XII, Chapter 152, §152.045, §152.052; Title XVIII (Public Health), Chapter 224 (Environmental Protection), §224.217, §224.218, §224.866, §224.895.

Louisiana

- As of April 1996, Louisiana's recycling rate was 6 percent.
- In 1986, Louisiana created a Litter Control and Recycling Commission that is responsible for promoting recycling and for educating the public about litter problems and recycling.
- The Department of Environmental Quality is directed "to conserve and recycle our natural resources. . . . maximum resource recovery from solid waste and maximum recycling and reuse of such resources must be considered goals of the state."
- State agencies are directed to use degradable or recyclable plastics whenever possible. The Department of Environmental Quality is responsible for providing a plan for the use of such plastics.
- In 1989, state agencies were directed to adopt policies to promote the use of products made from recycled materials. At the beginning of such programs, no less than 5 percent of the total goods ordered had to be recycled; the goal in five years was to reach a minimum of 25 percent.
- Further, the Department of Transportation was advised to initiate procedures to ensure that it used the maximum amount of recycled materials possible in highway maintenance and construction.
- Lead-acid batteries have been banned from land disposal

since 1989 and are required to be recycled. Battery retailers must post signage to this effect.

- As of 1 July 1990, used oil was banned from disposal in landfills, and beginning on 1 January 1991, citizens were required to dispose of oil only at permitted used oil collection facilities.
- After 1 July 1990, white goods could no longer be disposed of and had to be recycled instead.

SOURCE: Louisiana Statutes Annotated: Title 25, Chapter 24 (Statewide Beautification), Part I (Louisiana Litter Control and Recycling Commission), §1103, §1105; Title 30, Subtitle II (Environmental Quality), Chapter 2 (Department of Environmental Quality), §2038, Chapter 9, §2184, Chapter 18 (Solid Waste Recycling and Reduction Law), §2415, §2417, §2420, §2421; Title 36, Chapter 22, Part III, §918.

Maine

- As of April 1996, Maine's recycling rate was 33 percent.
- Annual reports to the legislature are required, to advise that body of the state's efforts to purchase and promote supplies and materials composed in whole or in part of recycled materials.
- A Waste Reduction and Recycling Loan Fund has been established to aid projects aimed at reducing and/or recycling solid and hazardous wastes and promoting the reuse of postconsumer materials.
- Maine has required each municipality to review its procurement procedures and specifications to make sure there is no discrimination against any products containing recycled materials.
- In 1989, Maine set a goal to reduce its waste by increasing its recycling rate to 50 percent by 1994.
- The state's solid waste management plan was submitted to the legislature on 1 March 1990. The Office of Waste Reduction and Recycling was subsequently created, within the Maine Waste Management Agency, to provide technical and financial assistance programs to municipalities across the state. These programs covered recycling, composting, source separation, marketing of materials, etc.
- By 1 July 1990, the capitol complex of state buildings was required to institute a recycling program that handled, at a

minimum, office paper, corrugated cardboard, and returnable containers.

- By 1 January 1991, all other state agencies were required to have implemented source-separated recycling programs and to have established and implemented a waste reduction program.
- Similar requirements for recycling programs have been set for private businesses within the state. By 1 July 1993, all businesses employing 15 or more persons at one site must have office paper recycling programs implemented.
- Tax incentives for recycling in Maine consist of a tax credit equal to 30 percent of the cost of equipment and machinery. Subsidies are also available to municipalities for the costs of transporting scrap metal.
- Maine prohibits the disposal and requires the recycling of lead-acid batteries.
- In 1995, the Maine Waste Management Agency was closed, and some of its responsibilities were transferred to the State Planning Office.

SOURCE: Maine Revised Statutes Annotated: Title 5, Part 4, Chapter 155 (Purchases), Subchapter I, §1812-A; Title 10, Part 2, Chapter 110, Subchapter II, §1023-G; Title 30A, Part 2, Subpart 9, Chapter 223, Subchapter I, §5656; Title 38, Chapter 24 (Maine Waste Management Agency), Subchapter II, §2122, Subchapter III (Office of Waste Reduction and Recycling), §2131, §2133, §2137, §2138.

Maryland

- As of April 1996, Maryland's recycling rate was 27 percent.
- The Maryland Recycling Act, passed in 1988, included the following requirements: (1) the creation of an Office of Waste Minimization and Recycling (OWMR); (2) the creation and implementation of a recycling plan by the OWMR to reduce waste generated by the state government by 20 percent (or at least 10 percent); (3) biannual reports to the governor and general assembly on recycling efforts, programs, and markets, beginning 1 January 1990; and (4) the creation and implementation of public education and grant programs. This law required the city of Baltimore and the larger counties (populations over 150,000) to recycle 20 percent of their solid waste, and smaller counties to recycle 15 percent of their waste by 1994.

- All counties, as well as the state government, were required to submit recycling plans in 1990 and to implement them by 1 January 1992.
- Maryland has recycled-content requirements for newsprint consumers, stated by the percentage of total newsprint used within a calendar year by a given publisher: 12 percent for 1992 and 1993, 20 percent for 1994, 25 percent for 1995, 30 percent for 1996, 35 percent for 1997, and 40 percent for 1998 and all subsequent calendar years.
- By 1 January 1991, all state agencies were to review their purchasing and procurement specifications and to require the use of a percentage price preference for the purchase of supplies and commodities containing recycled materials. The state government is required to purchase products with recycled content whenever possible, and a pricing preference of 5 percent is allowed. Government agencies responsible for maintenance of public lands are required to give preference to compost use when using public funds. By the year 2000, newspapers and telephone books distributed within the state must have a recycled content of 40 percent.
- The Maryland Used Oil Recycling Act bans the disposal of used oil and requires that it be collected and recycled. Signs explaining this law and giving the locations of government-operated used oil collection facilities must be posted at all locations where motor oils and lubricants are sold.
- The Used Tire Recycling Act of 1991 requires that tires be recycled. Additionally, a Used Tire Clean-Up and Recycling Fund was created in 1991 and assigned to the OWMR for management.
- After 1 July 1994, mercuric oxide battery manufacturers became responsible for the collection, transportation, and recycling or disposal of these batteries. Also, by 1 January 1995, each cell, rechargeable battery, or rechargeable product sold in the state was required to be covered by a collection, recycling, or disposal program, to be put in place by the marketer.
- As of October 1994, yard waste collected separately was banned from disposal facilities. Counties are required to study the feasibility of composting, and composting was officially included under the definition of "recycling"; therefore, diversion of such waste could be counted toward recycling goals.

SOURCE: Annotated Code of Maryland, 1988: Environment Division, Title 9 (Water, Ice, and Sanitary Facilities), Subtitle 17 (Office of Recycling), §9-1702, §9-1703, §9-1706, §9-1707; Natural Resources Division, Title 8 (Water and Water Resources), Subtitle 14 (Used Oil Recycling), §8-1401, §8-1411.1; State Finance and Procurement Division, Division II, Title 14 (Preferences), Subtitle 4, §14-405. Title 13 deals with vehicle laws and has a number of sections regulating scrap metals recycling with regard to vehicles.

Massachusetts

- As of April 1996, Massachusetts' recycling rate was 31 percent.
- The state empowers local agencies to establish and fund recycling programs. Such programs may be declared mandatory for residences, schools, and businesses, and source separation may be required.
- The state's nonmandatory integrated waste management goals for the year 2000, intended to treat the solid waste stream, call for source reduction to handle 10 percent of such waste, recycling to handle 38 percent, composting to handle 3 percent, and incineration/landfilled ash to handle 49 percent.
- In 1981, the state passed a mandatory deposit law on all beverage containers.
- An executive order in 1987 advised the state to buy products made from recycled materials whenever possible.

SOURCE: Massachusetts General Laws Annotated: Part I, Title VII, Chapter 40, §8H (Recycling Programs).

Michigan

- As of April 1996, Michigan's recycling rate was 25 percent.
- In 1976, Michigan passed a mandatory beverage container deposit law.
- The state has a recycling goal of 50 percent by 2005.
- State percentage requirements for the purchase of supplies and equipment containing recycled materials were set as follows: 10 percent for 1989, 15 percent for 1990, and 20 percent for 1991 and subsequent years.
- Annual reports to the governor and the legislature are required, detailing the state's procurement and purchasing procedures and accomplishments with regard to recycled materials.
- Goals for the purchase of paper products made from recycled paper and the percentage of postconsumer waste contained

in that paper have been set: In 1989, 30 percent of the paper purchased was required to be recycled and the minimum amount of postconsumer waste required was 25 percent; in 1990, 40 percent recycled, with a 35 percent minimum in postconsumer waste; and in 1991 and subsequent years, 50 percent recycled and a 50 percent minimum in postconsumer waste.

- County commissioners may impose a $2 per month or $25 per year surcharge per household to fund waste reduction, recycling, composting, and household hazardous waste programs.
- The Clean Michigan Fund Act created a commission to study the feasibility of recycling and composting in 1986. Subsequently, a capital grant program and an operational grant program were established to fund recycling and composting projects within the state. More funding was allocated for the study and development of recycled materials markets.
- Michigan's Office Paper Recovery Act set the following goals for recycling rates in state offices: 1 January 1989—20 percent or more; 1 January 1990—25 percent or more; 1 January 1992—50 percent or more; and 1 January 2000—85 percent or more.
- In 1988, a Plastics Recycling Development Fund Consortium was established to study plastics recycling, packaging, and production, funded from state revenues. As of 1 January 1992, this group was disbanded by statute.
- State agencies are required to recycle all used motor oil.
- In 1994, Public Act 451 was amended to address the following issues: household hazardous waste (Part 111); plastic recycling codes (Part 161); plastic degradable containers (Part 163); used oil (Part 167); scrap tires (Part 169); and dry cell and lead-acid batteries (Part 171).

SOURCE: Michigan Compiled Laws Annotated: Chapter 18, Article 2, §18.1261a and b, Chapter 124, §124.508a, Chapter 299 (Natural Resources), Clean Michigan Fund Act, §299.378, §299.382, §299.384, §299.387, §299.389a, Solid Waste Management Act, §299.406, Office Paper Recovery Act, §299.463, Plastics Recycling Development Fund Act, §299.473, §299.475, §299.476, Chapter 319, Used Oil Recycling Act, §319.314.

Minnesota

- As of April 1996, Minnesota's recycling rate was 44 percent, the highest in the United States.

- Minnesota has a 10 percent source reduction goal, but after 31 December 1996, further state goals are to be established through county plans.
- All municipal solid waste haulers must be licensed by the city and/or county, and they are required to have a volume-based fee structure.
- In 1989, Minnesota established a recycling rate goal of 25 percent for counties outside the Minneapolis-St. Paul metropolitan area and 35 percent for counties within the metropolitan area by 31 December 1993.
- The following items are banned from disposal in state landfills: waste tires, source-separated recyclable materials, lead-acid batteries, used oil, yard and tree waste, white goods, tax-exempt clothing, dry-cell batteries containing mercuric oxide electrodes, silver oxide electrodes, nickel-cadmium batteries, sealed lead-acid batteries, rechargeable batteries or battery packs, mercury, thermometers, thermostats, telephone directories, and assorted vehicle fluids.
- Public entities must comply with numerous procurement requirements regarding the purchase of recycled-content and recyclable products, double-sided copying, use of soy inks, and other such measures. A 10 percent purchasing incentive is allowed for the procurement of state commodities that contain recycled materials. Additional consideration is to be given to products made from waste generated within the state.
- The Department of Natural Resources is required to provide recycling containers at all state parks, and all public buildings are required to have recycling containers that collect at least three materials.
- Technical assistance and funding are available for the development of recycling programs, facilities, and markets.
- All counties are mandated to give residents the opportunity to recycle. The following must be available in each county: at least one recycling center, convenient sites for the collection of recyclables, and public education information and promotion programs. Furthermore, cities with populations of 5,000 or more must provide curbside pickup, centralized drop-off, or a local recycling center for at least four kinds of recyclables.
- All materials collected for recycling must be taken to markets for sale or reprocessing.
- To be designated a recycling center, a facility must be open a minimum of 12 hours each week, 12 months of the year, and it must accept at least four different materials.

- All containers, receptacles, and storage bins must be partly made from recycled materials and be recyclable themselves, if at all possible.
- State funding is available to counties for developing and implementing programs on source reduction, recycling, and recycling markets.

SOURCE: Minnesota Statutes Annotated: Chapter 16B, §16B.121, Chapter 85 (Division of Parks and Recreation), §85.205, Chapter 88, §88.80, Chapter 115A (Waste Management), §115A.48, §115A.551–553, §115A.555–557, §115A.904, §115A.915, §115A.916, §115A.931, §115A.95, §115A.9561, Chapter 325E, §325E.115, Chapter 473, §473.8441. Bill Dunn, Policy Analyst, Minnesota Pollution Control Agency.

Mississippi

- As of April 1996, Mississippi's recycling rate was 8.5 percent.
- This state promotes recycling and waste reduction through legislative declarations and has funding available to assist in recycling and solid waste reduction projects.
- Within 15 months after the publication of Subtitle D in the *Federal Register* (which occurred in October 1991), a nonhazardous solid waste management plan must be developed and administered. Such a plan is expected to include a 25 percent waste minimization goal and to define the strategy for attaining that goal by 1 January 1996. Subsequent to the passage of this bill, the state delegated this goal to each of its individual counties. As of 1994, all counties had submitted these waste management plans. Counties are further required to update these plans and submit them to the state every five years. A study is currently in progress to evaluate the state's waste reduction goals and the methods to achieve them. The study is expected to be submitted to the State Legislature by the end of 1997.
- Lead-acid batteries may not be disposed of and must be recycled. All sellers of such batteries must erect signage stating this law and offer collection facilities.
- A state agency wastepaper recycling plan was to be implemented by 1 July 1987. Since that time, progessive implementation of this plan has been, and continues to be, an ongoing project among individual state agencies.

- In 1990, a Commission on Environmental Quality was authorized and empowered to promote minimization, recycling, reuse, and treatment of wastes in lieu of disposal.
- From 2 April 1990 until 1 July 1992, a moratorium on permits for new or expanded nonhazardous waste facilities was instituted.
- By 1 July 1991, the Department of Economic and Community Development was expected to assist and promote the recycling industry, as well as create a Recycling Market Development Council. This council was directed to issue a comprehensive report, due 1 January 1993, on the state of the recycling industry and how best to promote it and source reduction within the state. A subsequent directive had the Council do another report on promoting and following through with ideas from the first report, to be presented in 1996. After presenting such reports to the governor in 1996, the Council was disbanded. (The Pollution Prevention Program is still required to report on the status of pollution prevention and recycling every five years.)
- By 1 July 1992, all state agencies, branches, and institutions must establish recycling and source reduction programs, including both collection of recyclables and procurement of recycled materials.
- Other statutory requirements direct the state and the Mississippi Multimedia Pollution Program as follows:

  —To promote pollution prevention, reuse, and recycling of waste in lieu of treatment or disposal of waste;
  —To require all state agencies to aid and promote the development of recycling through the establishment of recycling programs as well as aiding and promoting the development of recycling markets through the establishment of policies for the procurement of goods containing recycled materials;
  —To compile, organize, and make available information on pollution prevention and recycling technologies and procedures;
  —To sponsor and conduct workshops and conferences on pollution prevention and recycling.

- A 1991 Waste Tire Law requires a $1 waste tire fee to be charged on each new tire sold; the monies from this program go to a grant fund for the Solid Waste Program and are specifically earmarked for waste tire projects and programs.

SOURCE: Mississippi Code 1972 Annotated: Title 17, Chapter 17 (Solid Wastes Disposal), §17-17-33, §17-17-41, §17-17-59, §17-17-101, §17-17-103, §17-17-123, Nonhazardous Solid Waste Planning Act of 1991, §17-17-221, §17-17-227, Mississippi Regional Solid Waste Management Authority Act, §17-17-305, §17-17-319, §17.17-401, §17-17-403, §17-17-429, §17-17-431; Title 29, Chapter 5, §29-5-2; Title 31, Chapter 7, §31-7-13; Title 49 (Conservation and Ecology), Chapter 31, Mississippi Comprehensive Multimedia Waste Minimization Act of 1990, §49-31-5, §49-31-7 (as amended in 1994), §49-31-9, §49-31-11, §49-31-15, §49-31-17.

Missouri

- Missouri's recycling rate was 18 percent in April 1996.
- Purchasing specifications in this state must eliminate any discrimination against the procurement of supplies and goods made from recycled materials, and they are expected to reflect a policy of reducing and ultimately eliminating products manufactured in whole or in part from polystyrene foam that uses any fully halogenated chlorofluorocarbons.
- The required minimum percentages of recycled materials in state-purchased paper products are specified as follows: 40 percent for newsprint, 80 percent for paperboard, 50 percent for high-grade printing and writing paper, and 5 to 40 percent in tissue products. These goals are to be attained in percentages of total state purchases by the following dates: 10 percent in 1991 and 1992, 25 percent in 1993 and 1994, 40 percent in 1995, and 60 percent by 2000.
- Additionally, all state agencies are required to recycle, at a minimum, 25 percent of the paper and paper products used and 75 percent of all used motor oil.
- After 1 January 1990, all major counties were required to ensure that, to the greatest extent possible, all recyclable and reusable materials were removed from the waste stream prior to disposal or incineration.
- By 1 July 1990, all state agencies were expected to source separate and recycle plastics, paper, metals, and other recyclable items.
- The Department of Natural Resources (DNR) is expected to conduct and contract for research into alternatives to the landfill disposal of solid wastes. It is also directed to prepare model solid waste management plans for rural and urban areas that emphasize waste reduction and recycling and are designed

to achieve a reduction of 40 percent, by weight, in solid waste disposed of by 1 January 1998. Promotion of resource recovery through education, market development, technical assistance, and the establishment of state government recycling programs is another of the department's duties.

- After 1 January 1991, the following items are prohibited from disposal in landfills and are required to be collected for recycling: white goods, waste oil, and lead-acid batteries. Signs regarding the battery disposal and recycling law are to be posted at all points of sales for lead-acid batteries.
- After 1 January 1991, waste oil may not be burned without energy recovery.
- After 1 January 1992, yard waste is banned from landfills.
- After 1 January 1994, all major newspaper publishers must file annual statements certifying the number of tons of newsprint they have used in the past year and the average recycled content of that newsprint (see statute for specific target percentages).
- Missouri requires that all plastic products, bottles, and rigid containers must be labeled with identifying resin codes (per Society of the Plastics Industry specifications).
- Transfer stations with the sole function of separating materials for recycling are exempt from landfill fees.
- A total of $1 million has been appropriated from the solid waste management fund for fiscal years 1992–1997 for activities that promote the development of markets for recycled materials.
- The DNR is directed to collect and disseminate information, as well as to conduct educational and training programs that promote resource recovery and recycling.
- A Source Reduction Advisory Board was to be established for one year to research and write a report that would include information on how to reduce the amount of packaging material in the waste stream as well as maximize the recycling and reuse of packaging materials.
- The duties of the Missouri Rural Economic Development Council include assisting existing businesses and encouraging new businesses that promote resource recovery, waste minimization, and recycling.

SOURCE: Vernon's Annotated Missouri Statutes: Title IV, Chapter 34, §34.031, §34.032, Chapter 37, §37.078; Title XVI (Conservation, Resources, and Development), Chapter 260 (Environmental Control), §260.005, §260.035, §260.200, §260.201, §260.202, §260.209, §260.210, §260.225, §260.248, §260.250, §260.255, §260.260, §260.262, §260.281, §260.310, §260.320, §260.325, §260.330, §260.335, §260.342, §260.344, §260.345; Title XL, Chapter 620, §620.157, §620.161, §620.605.

Montana

- As of April 1996, Montana's recycling rate was 6 percent.
- A 25 percent tax credit is available for capital investments in equipment used for collecting or processing recyclable materials or for manufacturing products from such materials. This provision terminated 31 December 1995.
- An additional 5 percent tax deduction is available for taxpayers who purchase recycled materials as business-related expenses. This provision terminated 31 December 1995.
- The Department of Motor Vehicles specifically encourages the recycling of used or outdated vehicle license plates, which are made from aluminum.
- The legislature has stipulated that "it is the continuing responsibility of the state of Montana to use all practicable means . . . to improve and coordinate state plans . . . to the end that the state may enhance the quality of renewable resources and approach the maximum attainable recycling of depletable resources." A public policy articulated elsewhere declares that "maximum recycling from solid waste is necessary to protect the public health, welfare, and quality of the natural environment."
- As of 1 January 1997, the Pollution Prevention Bureau was assigned the duty of acting as "a clearinghouse for information on waste reduction and reuse, recycling technology and markets, composting, and household hazardous waste disposal, including chemical compatibility." The bureau is also expected to oversee the state's waste management and resource recovery plan. Further, the bureau coordinates a statewide household hazardous waste public education program.
- Another state policy declares that it must plan for and implement an integrated approach to solid waste management, in the following order of priority: source reduction, reuse, recycling, composting, and, finally, landfill disposal or incineration. As part of this policy, each state agency, the legislature, and the university system were required to prepare source reduction and recycling plans by 1 January 1992; these plans

had to include provisions for composting yard wastes and recycling office and computer paper, cardboard, used motor oil, and any other recyclables for which a market could be found or developed. Source reduction and recycling programs were established and implemented on 1 July 1992.

- As of 1 January 1992, state purchasing specifications promote products made from recycled materials, including, at a minimum: paper, paper products, plastic, plastic products, glass, glass products, tires, and motor oil and lubricants.
- The state goal was to have 95 percent of the paper and paper products used by state agencies, universities, and the legislature made from recycled materials that maximize postconsumer material content by 1 January 1996. "Substantial steps" have been made toward use of recycled office material, but actual percentages have not been measured, according to Peggy Nelson, Montana's Waste Reduction and Recycling Coordinator.
- Waste oil may not be disposed of and must be recycled. All sellers of motor oil must post this information and give the location of the nearest waste oil collection center within 25 miles.

SOURCE: Montana Code Annotated: Title 15, Chapter 32, Part 6 (Recycling of Material), §15-32-601 through 605, §15-32-610 and 611; Title 61 (Motor Vehicles), Chapter 3, Part 3, §61-3-336; Title 75 (Environmental Protection), Chapter 1, Part 1, §75-1-103; Chapter 10 (Waste and Litter Control), Part 1, §75-10-101 through 104, §75-10-121, Part 2, §75-10-203, §75-10-215, Part 5 (which has extensive laws dealing with the recycling and disposal of motor vehicles, not described above), Part 8 (Integrated Waste Management), §75-10-802, §75-10-804 through 807, Part 11, §75-10-1101.

Nebraska

- The legislature has declared that a comprehensive statewide litter and recycling program is necessary. A Nebraska Litter Reduction and Recycling Fund (to be funded by littering fees) and a Waste Reduction and Recycling Incentive Fund were created to assist in implementing this program. As of 1 July 1991, an annual waste reduction and recycling fee was imposed on all state retail businesses, with all such fees credited to the Waste Reduction and Recycling Incentive Fund; proceeds from fees imposed on new tire sales are also credited to this fund.

SOURCE: Nebraska Revised Statutes of 1943: Chapter 69, Article 20, §69-2006; Chapter 77, Article 23, §77-2368; Chapter 81, Article 15B (Environmental Protection—Litter Reduction and Recycling), §81-1535, §81-1545, §81-1546, §81-1558, §81-1561, Article 15L (Waste Reduction and Recycling), §81-15,160, §81-15,161, §81-15,163, §81-15,165, Article 16F (State Government Recycling Management Act), §81-1646.

Nevada

- As of April 1996, Nevada's recycling rate was 12 percent.
- Nevada has no legislation at this time that deals with recycling, other than the recycling of wastewater and abandoned vehicles.

SOURCE: Nevada Revised Statutes: Title 40, Chapter 445; Title 43, Chapter 487.

New Hampshire

- The state's general court has declared that the goal of the state for the years 1990–2000 is to reduce the weight of the solid waste stream by a minimum of 40 percent.
- Integrated waste management solutions are supported by the general court of the state in the following priorities: (1) source reduction; (2) recycling, reuse, and composting; (3) waste-to-energy technologies; (4) incineration (without resource recovery); and (5) landfilling.
- State agencies are required to use products incorporating recycled materials to the maximum extent possible, and information on the percentage of postconsumer waste must be included by all vendors of such products.
- A 5 percent price incentive is allowed for the state's purchase of recycled paper products. This incentive was to be applied to products containing the following percentages of recycled fiber: 10 percent by June 1991, 25 percent by June 1992, and 40 percent by June 1996.
- These percentage requirements are also applicable to the goals for the amounts of recycled paper to be purchased (e.g., 10 percent of the total amount of paper purchased by the state had to be recycled paper by June 1991).

- All state agencies are required to have recycling programs for postconsumer wastes that they generate and for which markets have been identified.
- The state's policies include implementing steps to encourage reduction of packaging and the maximum possible use of products made from recycled materials by state agencies. Any procurement specifications that discriminate against recycled materials are prohibited.
- As of 31 December 1990, all state agencies were to have implemented waste reduction and recycling programs.
- A Waste Management Council was established in 1990.
- The Division of Waste Management's responsibilities include promoting the implementation of recycling technologies as well as identifying and establishing recyclable materials markets. The division is also directed to develop and implement technical assistance programs and to allocate funds for solid waste management programs to include recycling.
- Each town in the state is responsible for its own solid waste management plans, which must consider environmental impact, economic impact (including resource recovery and recycling), and area impact.
- As of 1 January 1992, "no person shall dispose of refuse at any private solid waste landfill facility having a lining and leachate collection system, unless all recyclable materials have been removed from such refuse, or such refuse has been otherwise reduced in weight by at least 20 percent."
- Wet-cell batteries may not be disposed of and are required to be recycled.
- A recycled/recyclable labeling emblem or logo has been established by the Department of Environmental Services and may only be used as the department deems appropriate. Unauthorized use of the emblem can result in a fine of $2,500 per day for each day on which the violation continues.
- Used oil must be recycled, and the recycling of lead-acid batteries and used tires is encouraged.

SOURCE: New Hampshire Statutes Annotated: Title I, Chapter 6, §6:12, Chapter 21-I, §21-I:1-a, §21-I:11, §21-I:14-a, §21-I:60 through 21-I:65, Chapter 21-O (Department of Environmental Services), §21-O:9; Title X, Chapter 147, §147:43, Chapter 149-M (Solid Waste Management), §149-M:1, §149-M:1-a, §149-M:3, §149-M:8, §149-M:10, §149-M:13-a and b, §149-M:17 through 149-M:19, §149-M:22, §149-M:24, §149-M:25, §149-M:31, Chapter 149-N (Recycling Logo).

New Jersey

- As of April 1996, New Jersey's recycling rate was 42 percent, the second highest in the country.
- As of 1987, this state mandated the source separation and collection of recyclables. A goal was set to recycle at least 25 percent of the solid waste stream by 1992.
- A New Jersey Office of Recycling was established within the Department of Environmental Protection in 1987, but most of the responsibility for recycling is assigned to individual counties. These counties are responsible for administering their own recycling plans, which must provide for the collection and recycling of at least three materials in addition to leaves.
- A recycling tax of $1.50 per ton of solid waste accepted for disposal is applied to all solid waste facilities, and the monies are credited to the State Recycling Fund, which is a funding source for recycling programs throughout the state. A Statewide Mandatory Source Separation and Recycling Program Fund has also been established to aid counties and municipalities in implementing their recycling programs, as well as for the study of recyclable materials markets.
- Plastic and bimetal beverage containers may not carry a recyclable emblem unless the state determines that feasible recycling systems for those products are available.
- Leaves may not be disposed of in landfills.
- State procurement guidelines prohibit discrimination against recycled materials, and a bidding preference is given for recycled paper containing the highest amount of postconsumer waste. Goals for the purchase of recycled paper or recycled paper products, as a percentage of the total amounts of such products purchased, were set as follows: 10 percent by 1 July 1987, 30 percent by 1 July 1988, and 45 percent by 1 July 1989.
- State specifications encourage the maximum usage of such products as recyclable asphalt pavement, crumb rubber, and glass aggregate paving materials. All products made from recycled materials must be given preference whenever the price is "reasonably competitive."
- A 50 percent taxpayer credit is available for purchasers of recycling equipment. Receipts from the sales of recycling equipment are exempt from the sales and use tax.

SOURCE: New Jersey Statutes Annotated: Title 13 (The New Jersey Statewide Source Separation and Recycling Act), Chapter 1D

(Department of Environmental Protection), §13:1D-18.3, Chapter 1E (Solid Waste Management), §13:1E-95, §13:1E-96, §13:1E-96.1, §13:1E-99.13 through 13:1E-99.21, §13:1E-99.24 through 13:1E-99.28, §13:1E-99.30, §13:1E-99.31, §13:1E-99.33, §13:1E-99.34, §13:1E-99.37, §13:1E-99.38; Title 52, Subtitle 5, Chapter 34 (Purchase of Recycled Materials), §52:34-21 through 52:34-24; Title 54 (Taxation), Subtitle 4, Part 1, Chapter 10A, §54:10A-5.3, Subtitle 4A, Chapter 32B, §54:32B-8.36.

New Mexico

- As of April 1996, New Mexico's recycling rate was 12 percent.
- The state's Litter Control and Beautification Act declares a need for a state-coordinated plan of education, control, prevention, and recycling in order to eliminate litter. The Litter Control Council must have a member from RECYCLE New Mexico.
- The Department of Highways is required to encourage voluntary recycling programs, aid in identifying programs and available markets for recycled materials, and promote private recycling efforts for all recyclable items.
- In 1990, the state passed a Solid Waste Act. One of its purposes is to "plan for and regulate, in the most economically feasible, cost-effective and environmentally safe manner, the reduction, storage, collection, transportation, separation, processing, recycling, and disposal of solid waste." Another purpose of the act is to require that all state agencies set procurement policies to aid and promote the development of recycling as well as markets for recyclable materials.
- By 1 July 1992, each state agency and the legislature was to implement both a source reduction program and a source-separated recycling program. The recycling program had to include, at a minimum, high-grade paper, corrugated paper, and glass. Postsecondary educational institutions were also directed to implement similar programs by this date, and they were additionally directed to include a composting component.
- By 31 December 1992, a comprehensive and integrated solid waste management program was submitted to the Environmental Improvement Board. The following priorities were identified: (1) source reduction and recycling, (2) environmen-

tally safe transformation, and (3) environmentally safe landfill disposal. Other specific components of the plan now include public education programs on solid waste management, recycling, source reduction, and composting. This solid waste program was fully implemented on 1 July 1994. At least once every three years, it is to be reexamined for management, efficiency, and compliance.

- By 1 July 1994 and annually thereafter, the director of the Department of Environmental Improvement is to prepare a report for the legislature on the status of solid waste management efforts in the state.
- A Division of Solid Waste was created by the Solid Waste Act to enforce and implement the act and its programs. Its responsibilities include providing technical assistance on solid waste issues, promoting and planning source reduction and recycling programs and facilities, developing information for education and technical assistance, and researching and developing recycling markets.
- The state purchasing agent and all procurement personnel are directed to establish specifications to promote goods and supplies containing recycled materials. A 5 percent price incentive is offered for such goods.
- Tax incentives in the form of investment credits are offered for equipment used in recycling or recycling/reprocessing facilities.
- The Tire Recycling Act of 1994 provided the Department of Environmental Improvement with "the direction for the development of programs to address scrap tire recycling, disposal, and funding assistance for local governments." It also required that as of 1 July 1994, the Department of Motor Vehicles was to assess each motorcycle registration applicant 50 cents and passenger car registration applicants $1, to supply the Tire Recycling Fund. (Varying fees are also collected on bus and truck registrations.)

SOURCE: New Mexico Statutes 1978, Annotated: Chapter 7 (Taxation), Article 9A (Investment Credit); Chapter 67, Article 16 (Litter Control and Beautification), §67-16-2 through 67-16-4, §67-16-12; Chapter 74 (Environmental Improvement), Article 9 (Solid Waste Act), §74-9-2 through 74-9-4, §74-9-6, §74-9-10, §74-9-12 through 74- 9-17, §74-9-19.

New York

- As of April 1996, New York's recycling rate was 32 percent.
- This state's overall goal is to attain a solid waste reduction of 50 percent through integrated efforts by 1997.
- All beverage containers sold within the state are subject to a mandatory deposit.
- New York's Department of Energy Conservation (DEC) must serve as a clearinghouse for information pertaining to the reduction and recycling of waste generated by commercial and industrial enterprises. The DEC is expected to provide consumer education on the "economic and environmental benefits of solid waste management practices and the concomitant needs for waste reduction and for consumers to actively seek consumer products which contain secondary materials or which are easily recycled or reused." An annual report on solid waste and progressive reduction efforts is to be submitted to the governor.
- The Department of Economic Development and the DEC are directed to work together to conduct secondary materials market development programs.
- The DEC is responsible for designing and implementing a local resource reuse and development program that must promote the collection, intermediate processing, and marketing of waste materials that were formerly disposed of as municipal solid waste within the state.
- Within the DEC, a Bureau of Waste Reduction and Recycling was established to assist in the development and promotion of local waste reduction, source separation, and recycling programs. The bureau was directed to establish a recycling emblem and set the standards for its use, as well as penalties for its misuse.
- Lead-acid batteries may not be disposed of and must be collected for recycling.
- State-funded assistance is available for municipal recycling projects and innovative recycling demonstration projects.
- Municipalities are empowered by the legislature to require source separation of recyclable or reusable materials. As of 1 September 1992, this requirement is to be adopted into local laws and ordinances to cover, at a minimum, the recycling of paper, glass, metals, plastics, and garden and yard waste.
- All state and judicial agencies are required to buy recycled products, within a price incentive of 10 percent; the limit is

raised to 15 percent if the product is made from at least 50 percent postconsumer waste.

- All public utilities are required to recycle industrial materials, as well as to use supplies and goods incorporating recycled materials whenever possible.

SOURCE: McKinney's Consolidated Laws of New York Annotated: State Solid Waste Management Act of 1988, Chapter 15, Article 14 (Energy Conservation Services), §263, Article 27; Title 4 (Marketing of Recyclable Materials), §27-0401, §27-0403, §27-0405; Title 7 (Solid Waste Management and Resource Recovery Facilities), §27-0717, Chapter 24, Article 6 (Public Health and Safety), §120-aa, Chapter 30, Article 2-A, §40-a, Chapter 43-A, Article 5; Title 12, §1285-g (Industrial Materials Recycling Program), Chapter 43-B, Article 27; Title 17 (Lead-Acid Battery Recycling), §27-1701, Article 54 (Implementation of the 21st Century Environmental Quality Bond Act); Title 6 (Innovative Recycling Demonstration and Municipal Recycling Projects), §54-0601, §54-0603, §54-0605, §54-0607, §54-0609, §54-0611.

North Carolina

- As of April 1996, North Carolina's recycling rate was 30 percent.
- North Carolina's overall goal was to recycle 25 percent of its total waste stream by 1 January 1993. When the general assembly decided on this figure in 1989, it also stated that it would evaluate the process of reaching that goal and would consider increasing the goal "as appropriate." The general assembly had arrived at this decision after stating that the state's failure to economically recover resources and energy from solid waste had resulted in unnecessary waste and depletion of natural resources. For these reasons, one of the state's goals is the maximum resource recovery from solid waste and maximum recycling and reuse of such resources.
- North Carolina provides tax deductions for the purchase and installation of recycling or resource recovery equipment.
- The Pollution Prevention Pays Program is a nonregulatory program that was established by statute in 1989 to encourage voluntary waste and pollution reduction efforts through research and the provision of "information, technical assistance, and matching grants to businesses and industries

interested in establishing or enhancing activities to prevent, reduce, or recycle waste."

- When the Division of Solid Waste Management was created within the state's Department of Health in 1989, it was to "promote sanitary processing, treatment, disposal, and statewide management of solid waste and the greatest possible recycling and recovery of resources." This was to be accomplished through public education, program development, and market development created to promote recycling.
- The Department of Health is also required to maintain a directory of state recycling and resource recovery systems. Comprehensive solid waste management reports must be prepared each year, and recommendations concerning all aspects of recycling and resource recovery are to be included.
- The state subscribes to the integrated solid waste management hierarchy in the following descending order of preference: (1) source reduction, (2) recycling/reuse, (3) composting, (4) waste-to-energy incineration, (5) volume-reduction incineration, and (6) landfill disposal.
- All counties, either individually or in a cooperative effort with others, were required to develop comprehensive solid waste management plans that included provisions to address the goal of recycling 25 percent of the total waste stream by 1 January 1993. Counties, cities, and towns were permitted to require participation in recycling programs.
- By 1 July 1991, all designated local governments were to initiate recycling programs.
- All beverage containers with detachable metal rings or tabs were prohibited as of 1 January 1990.
- All packing or packaging materials made from fully halogenated chlorofluorocarbons were prohibited after 1 October 1991.
- Plastic bags offered by retail establishments after 1 January 1991 must be recyclable and imprinted with such information. After 1 January 1993, the state must certify that at least 25 percent of these plastic bags are actually being recycled by individual retail establishments; if not, the retail establishment is prohibited from distributing them.
- Polystyrene food packaging may not be sold or distributed after 1 October 1991 unless it is recyclable. After 1 October 1993, the state must certify that at least 25 percent of such packaging is actually being recycled; if not, the packaging may no longer be sold or distributed.

- As of 1 July 1991, every plastic container had to carry a molded label identifying the resin from which it was made (per the Society of the Plastics Industry specifications).
- Products banned from landfills and required to be recycled as of the date indicated include: used tires, 1 March 1990; used oil, 1 October 1990; lead-acid batteries, 1 January 1991; white goods, 1 January 1991; and yard trash, 1 January 1993.
- A Solid Waste Management Trust Fund is administered by the Department of Health. One of its purposes is to provide funding for activities that promote waste reduction and recycling. Another funding source in the statutes is the North Carolina Solid Waste Management Loan Program.
- By 1 January 1992, all state government agencies, the general assembly, the general court, and the state university were required to establish recycling programs to (1) collect, at a minimum, aluminum, high-grade office paper, and corrugated paper; (2) find markets for their materials; and (3) encourage source reduction.
- All state agencies are required to use composted products whenever possible.
- All school boards are required to encourage recycling and develop curriculum materials and resource guides for recycling awareness programs.
- The Department of Transportation is encouraged to use recycled materials whenever possible.

SOURCE: General Statutes of North Carolina: Chapter 105 (Taxation), Subchapter I, Article 3, Schedule C, §105-122, Article 4, Schedule D, Division I, §105-130.5, §105-130.10, Subchapter II, Article 12, §105-275, Chapter 113 (Conservation and Development), Subchapter I, Article 1, §113-8.01; Chapter 130A (Public Health), Article 9 (Solid Waste Management), Part 1, §130A-290, Part 2, §130A-291, §130A-294, §130A-309.01 (Solid Waste Management Act of 1989), Part 2A, §130A-309.03 through 130A-309.10, §130A-309.12, §130A-309.14 through 130A-309.17, §130A-309.19, §130A-309.20, §130A-309.24, §130A-309.25, §130A-309.28, §130A-309.52; Chapter 136 (Roads and Highways), Article 2, §136-28.8, Chapter 143, Article 8, §143-129.2, Chapter 143B (Executive Organization Act of 1973), Article 7 (Department of Environment, Health, and Natural Resources), Part 4A (Governor's Waste Management Board), §143-285.10; Chapter 153A (Counties), Article 6, §153A-136, Article 22, §153A-422, §153A-427; Chapter 159I, §159I-2, §159I-8, §159I-11; Chapter 160A (Cities and Towns), Article 8, §160A-192.

North Dakota

- As of April 1996, North Dakota's recycling rate was 22 percent.
- This state's Solid Waste Management and Land Protection Act includes goals of promoting resource recovery and both promoting and assisting in the development of recycled materials markets. The act also declares that the state goal is to "encourage by 1995 at least a 10 percent reduction in volume of municipal waste deposited in landfills, by 1997 at least a 25 percent reduction, and by 2000 at least a 40 percent reduction."
- A Solid Waste Management Fund was established to help finance resource recovery and recycling projects that emphasize market development.
- Public schools are encouraged to develop and disseminate educational materials promoting municipal waste reduction, source separation, reuse, and recycling.
- All state agencies are encouraged to specify the use of soybean-based ink when purchasing newsprint printing services.
- By 1 July 1990, at least 15 percent of the garbage can liners purchased by state agencies must be starch based. That percentage must increase by 5 percent annually until at least a 50 percent level is reached.
- Beginning 1 July 1993, at least 10 percent of the total volume of paper and paper products purchased by state agencies must contain at least 25 percent recycled materials. This percentage increases as follows: 30 percent beginning 1 January 1994, 40 percent beginning 1 January 1996, 60 percent beginning 1 January 1998, and 80 percent beginning 1 January 2000.

SOURCE: North Dakota Century Code: Title 23 (Health and Safety), Chapters 23–29 (Solid Waste Management and Land Protection Act), §23-29-02, §23-29-03, §23-29-07.1, §23-29-07.5, §23-29-07.10; Title 54, Chapter 54-44.4 (State Purchasing Practices), §54-44.4-07, §54-44.4-08.

Ohio

- As of April 1996, Ohio's recycling rate was 31 percent.
- The state's goal was to recycle 25 percent of its solid waste by 1994.
- A Division of Litter Prevention and Recycling exists in the Department of Natural Resources. This division is required

to administer grants and to establish and implement a comprehensive statewide litter prevention and recycling program to include public education, technical assistance, identification of existing and potential markets for recycled materials and of any barriers to their use, and encouragement of state agencies to purchase products made from recycled materials. Within the division, there is a Litter Prevention and Recycling Advisory Council, as well as a Keep Ohio Beautiful Commission.

- Tax credits are available for cash donations to corporations whose sole purpose is to promote and encourage recycling. Additional taxes were also established in 1981 to help fund litter prevention and recycling in the state.
- As of 1 July 1982, all metal beverage containers with detachable pieces were prohibited.
- As of December 1993, yard waste may no longer be disposed of in landfills.

SOURCE: Page's Ohio Revised Code Annotated: Title 15 (Conservation of Natural Resources), Chapter 1502 (Litter Control and Recycling), §1502.01 through 1502.08, §1502.99; Title 57 (Taxation), Chapter 5733, §5733.064, §5733.066, §5733.122.

Oklahoma

- As of April 1996, Oklahoma's recycling rate was 12 percent.
- The Department of Health was directed by the legislature to provide, by 1 January 1991, a prioritized agenda of studies necessary to provide information that would facilitate the design and implementation of recycling initiatives.
- By 1992, the department was also directed to design and implement a statewide educational program on waste management that stressed recycling, litter reduction, waste reduction, and the management and disposal of "hard to dispose of" wastes.
- State agencies are required to participate in a recycling program, and a State Recycling Revolving Fund was created to assist in the funding of such recycling programs.
- Waste tires must be recycled, and a tax on new tires helps to fund this state's tire recycling program.

SOURCE: Oklahoma Statutes Annotated: Title 47, Chapter 62A (which contains a comprehensive Automotive Dismantlers and

Parts Recycler Act); Title 63 (Public Health and Safety), Chapter 1, Article 24, §1-2440 through 1-2443; Title 68 (Revenue and Taxation), Chapter 2 (Oklahoma Waste Tire Recycling Act), §53001 through 53008.

Oregon

- As of April 1995, Oregon's recycling rate was more than 30 percent.
- On 4 August 1983, the Oregon Recycling Opportunity Act was signed. Many consider this legislation to be the first mandatory recycling law ever passed on a statewide basis. The act required that studies be conducted to assess the state's various waste streams to identify all possible recyclables. Individuals within each "wasteshed" area would then be notified as to what could be recycled. In addition, by 1 July 1986, cities and counties were mandated to provide the "opportunity to recycle" to all residents, although citizen participation was not required. Source separation of recyclable items was also mandated, and each wasteshed had to identify and explain how this opportunity was being provided and give details on its public education and promotional programs. The act also states that after an unspecified period of time, a wasteshed may require that certain items be recycled.
- A collection service may charge people who do not source separate their recyclables higher fees than those who do.
- Oregon does have an antiscavenging law that prohibits the removal of recyclables from any container without the permission of the owner of the container or recyclables.
- It is illegal for anyone to mix source-separated recyclables with solid waste being collected for disposal.
- The highest percentage possible of all paper used by state agencies in any given year must be recycled paper. In Oregon, paper is considered recycled if not less than 50 percent of its total weight consists of secondary waste materials or not less than 25 percent of its total weight consists of postconsumer waste.
- All state agencies must ensure: (1) that their purchasing specifications do not discriminate against recycled products; (2) that they provide incentives for the maximum possible use of recycled materials and recovered resources; (3) that purchasing practices are developed to assure, as far as possible, that products bought are made of recycled materials or are

recyclable themselves; (4) that management practices are established to promote source reduction; and (5) that they use recycled paper wherever possible and require the same of all persons with whom they contract for services.

- Specific guidelines have been established to encourage paper conservation within state offices, and a 5 percent purchasing preference is given to products containing recycled materials.
- A 10 percent investment tax credit, now expired, was available for plastics recycling business investments made between 1 January 1986 and 1 July 1995.
- The Department of Environmental Quality is directed to establish percentages of plastic resins that must be recycled before a recycling program is deemed "effective." At least 15 percent of each plastic resin must be recycled statewide in 1992, but the required percentages may not exceed 75 percent before 31 December 1999.
- Oregon's legislature has established the following priorities for the state's integrated solid waste management system: (1) source reduction, (2) reuse, (3) recycling, (4) waste-to-energy incineration, and (5) landfill disposal or other disposal methods approved by the health department.
- The Department of Health is required to provide advisory technical and planning assistance for the development and implementation of recycling and solid waste management programs by state or local agencies or by anyone who provides solid waste collection services.
- In all statutes dealing with solid waste management, Oregon stresses the importance of waste reduction and recycling before landfill disposal. A $2 surcharge on each ton of solid waste disposed of in a landfill is collected to fund recycling programs.
- The state encourages both public education on household hazardous waste and the collection of such materials. The Department of Environmental Quality must implement statewide education programs on this subject.
- After 1 January 1990, lead-acid batteries were banned from landfills and required to be recycled. Signs with this information must be posted in all establishments that sell batteries; violators of this section are subject to a fine of $500 for each day that the sign is not posted.
- A fee may be charged on new tires, to be placed into a state-established Waste Tire Recycling Account. As of 1 July 1989, landfill disposal of whole tires was strongly discouraged but

not expressly prohibited. Anyone who purchases tire chips or waste tires for an approved reuse may receive a partial reimbursement of costs incurred in the reuse program.

- Used oil recycling is mandatory; other forms of disposal are prohibited, with certain exceptions.
- State agencies may not purchase any food packaging products that are not biodegradable or recyclable through a local program. Vendors who sell food items in state agencies are subject to the same requirement.
- Violators of the state's recycling laws can incur civil penalties of $500 per violation.
- Tax credits are available for costs incurred in providing a pollution control facility, which includes locations that recycle used oil (as well as other facilities that are primarily concerned with the prevention, control, or reduction of air, water, or noise pollution or of solid or hazardous waste).
- Waste haulers can qualify for waste diversion credits according to the amount of waste they can divert from landfills.
- Funding is available for recycling and resource recovery projects through both the Habitat Conservation Trust Fund and the Waste Reduction Trust Fund.

SOURCE: 1989 Oregon Revised Statutes: Title 26, Chapter 279 (Public Contracts and Purchasing), §279.729, §279.731, §279.733, §279.735, §279.737, §279.739; Title 29, Chapter 316 (Personal Income Tax Credits), §316.103, Chapter 317, §317.106; Title 36 (Public Health and Safety), Chapter 459 (Solid Waste Control), §459.005, §459.015, §459.035, §459.055, §459.165, §459.170, §459.175, §459.180, §459.185, §459.188, §459.190, §459.195, §459.200, §459.250, §459.280, §459.292, §459.294, §459.295, §459.305, §459.345, §459.411, §459.417, §459.420, §459.426, §459.614, §459.710, §459.770, §459.775, §459.780, §459.995, Chapter 468 (Pollution Control), §468.155, §468.165, §468.170, §468.190, §468.659, §468.660, §468.664, §468.675, §468.677, §468.679, §468.680, §468.681, §468.685, §468.850, §468.853, §468.856, §468.859, §468.862, §468.865, §468.869, §468.870, §468.871, §468.968, §468.969, Chapter 469 (Energy Conservation), §469.010, §469.185, §469.205.

Pennsylvania

- As of April 1996, Pennsylvania's recycling rate was 17 percent.
- The state had a recycling goal of 25 percent by 1 January 1997.

- The Municipal Waste Planning, Recycling, and Waste Reduction Act created funding for recycling programs in 1988 by imposing a recycling fee of $2 per ton for all solid waste processed at incineration facilities and most solid wastes disposed of at landfills. All monies generated from this fee are deposited in the Recycling Fund.
- After September 1990, all communities with populations greater than 10,000 must start recycling programs. By September 1991, communities with populations between 5,000 and 10,000 must do the same. Counties are responsible for creating and implementing solid waste disposal plans.
- A Recycling Advisory Committee was established within the Department of Environmental Resources (DER) to oversee the state's recycling program and to study other ways to encourage waste reduction. DER is also responsible for researching and identifying recycled materials markets.
- All municipalities must recycle at least three materials from the following group: glass; colored glass; aluminum; steel "tin" and bimetal cans; newsprint; high-grade office paper; plastics; and corrugated cardboard. Disposal facilities must also provide drop-off containers for at least three of these materials.
- All state and local agencies must recycle office paper, corrugated cardboard, aluminum, and leaf waste. Additionally, they are required to implement comprehensive waste reduction programs.
- All educational institutions must source separate and collect recyclable materials and institute source reduction programs as well.
- Lead-acid batteries must be recycled, and battery sellers cannot sell a new battery unless they receive the old one for recycling.
- Leaf waste is banned from landfills.
- Used oil is to be recycled.
- There is a procurement preference program for supplies and goods made from recycled materials. Of the total volume of paper purchased, at least 10 percent must be recycled paper in 1989, 25 percent in 1991, and 40 percent in 1993. Federal guidelines are to be followed with respect to recycled material content. Newsprint is considered recycled if it contains at least 40 percent postconsumer newspaper waste. Recycled paper must have a total weight consisting of not less than 20

percent secondary materials in 1989; not less than 30 percent secondary materials in 1991; not less than 40 percent secondary materials in 1993; and not less than 50 percent secondary materials in 1996 and thereafter, at least 10 percent of which must be postconsumer waste.
- All public agencies are expected to use composted materials as much as possible in maintaining public lands.

SOURCE: Purdon's Pennsylvania Statutes Annotated: Title 53, Part 1, Chapter 17A (The Municipal Waste Planning, Recycling, and Waste Reduction Act), Chapter 7 (Recycling Fee), §4000.701, §4000.702, §4000.706, Chapter 9 (Grants), §4000.902 through 4000.904, Chapter 15 (Recycling and Waste Reduction), §4000.1501 through 4000.1503, §4000.1509, §4000.1511; Title 58, Chapter 9 (Pennsylvania Used Oil Recycling Act), §476.

Rhode Island

- As of April 1996, Rhode Island's recycling rate was 24 percent.
- The state's recycling goal was 15 percent by 1993.
- Rhode Island passed the first mandatory source separation law in the country in June 1986: the Rhode Island Recycling Act.
- A Plastic Recycling and Litter Commission was created and required to submit a plan to the governor and general assembly by 1 January 1991 that would provide for the maximum recycling of plastic and foam food service items. For nonrecyclable products, the commission was to establish guidelines on photodegradability and biodegradability that would take effect on 1 January 1992.
- A Litter and Recycling Advisory Council was established to create recycling centers throughout the state and to encourage their use.
- By 1 January 1988, the state's central landfill was to construct and begin operating a recycling facility.
- Although the state recognizes that the definition of recyclable materials may change due to market conditions, it stipulates that the following materials must be separated and collected for recycling by all municipalities: food and beverage containers made of aluminum, glass, and other metals; newspapers; plastic soft drink bottles and milk jugs; and white goods.

- The Department of Environmental Management is directed to establish, maintain guidelines for, and enforce the proper use of a recycling logo.
- Only those beverage containers that have obtained a 50 percent recycling rate by 1992 and are free of any components that render them less recyclable or unrecyclable may be sold in Rhode Island.
- Telephone directories must be accepted for recycling by directory distributors, and the bindings must be such that they do not interfere with the recycling process.
- Commercial businesses have been required to separate and recycle office paper, corrugated cardboard, aluminum and steel "tin" cans, glass food and beverage containers, and newspapers since 1 January 1989. Businesses with more than 500 employees had to submit source reduction and recycling plans by 30 June 1989, those with more than 250 employees were required to do the same by 31 December 1989, and those with over 100 had to produce reports by 30 June 1990.
- By 30 June 1991, all state departments and agencies were to present recycling plans to the Department of Environmental Management.
- State officials are required to encourage the purchasing of recycled oil products.

SOURCE: General Laws of Rhode Island Annotated 1956, Reenactment of 1989: Title 21, Chapter 27.1 (Plastic Recycling and Litter Act), §21-27.1-3; Title 23, Chapter 18.8 (Waste Recycling), §23-18.8-3, §23-18.8-5, Chapter 18.12 (Beverage Container Recyclability), §23-18.12-3, Chapter 19 (Solid Waste Management Corporation), §23-19-31, Chapter 19.6, §23-19.6-8 and 6-9, Chapter 60, §23-60-7, Chapter 63, §23-63-4; Title 37, Chapter 15 (Litter Control and Recycling), §37-15-4; Title 42, Chapter 20, §42-20-16.

South Carolina

- As of April 1996, South Carolina's recycling rate was 16 percent.
- The South Carolina Solid Waste Policy and Management Act of 1991 directed the state to reach a goal of reducing the amount of solid waste received at landfills and incinerators by 30 percent (calculated by weight) by 27 May 1997. The act also created the goal of recycling at least 25 percent (by weight)

of the total waste stream generated in the state by 27 May 1997. A study is currently in progress to assess the ability and methods by which the state may attain these goals. This study is expected to be completed in 1998.

- On 27 November 1992, the Department of Health and Environmental Control submitted its first state solid waste management plan to the governor and general assembly. This plan included an inventory and evaluation of the state's solid waste situation and suggestions for recycling programs, technical and financial aid for local governments, and waste tire management. The plan is updated annually.
- The act also established both a Solid Waste Advisory Council and a Recycling Market Development Council.
- Disposal of lead-acid batteries and used oil in municipal solid waste landfills was prohibited as of 27 May 1992.
- Yard trash was banned from municipal solid waste landfills as of 27 May 1993.
- After 23 October 1993, whole waste tires were prohibited from landfill disposal.
- White goods were banned from landfills as of 27 May 1994.
- The act required that by 27 May 1996 at least 35 percent of the quantity of newspaper sold in the state be recycled.

SOURCE: South Carolina Department of Health and Environmental Control, Division of Solid Waste Management, *"An Overview of the S.C. Solid Waste Policy and Management Act of 1991,"* Columbia, SC: South Carolina Department of Health and Environmental Control, 1994.

South Dakota

- As of April 1996, South Dakota's recycling rate was 20 percent.
- Counties are directed to establish solid waste management policies.
- South Dakota subscribes to the following priorities of integrated waste management: (1) source reduction, (2) recycling and reuse, (3) incineration, and (4) landfill disposal.
- State agencies are to purchase recycled paper if the price and quality are competitive. Agencies using recycled paper may print the notation "printed on recycled paper" on any product that has been certified as such by the state.
- Procurement goals for recycled paper are set as percentages of the total volume of paper and paper products purchased

in a given fiscal year: at least 10 percent in 1992, at least 20 percent in 1993, at least 30 percent in 1994, at least 40 percent in 1995, and at least 50 percent for 1996 and subsequent years.
- Elementary and secondary schools are to attempt to recycle any used textbooks before disposing of them as solid waste.
- After 1 July 1988, all beverage containers offered for sale must be reusable, recyclable, or biodegradable.
- After 1 January 1991, no beverage containers may be offered for sale if they are connected to other beverage containers by a material that is neither biodegradable nor photodegradable.
- After 1 January 1992, no plastic garbage bag or garbage can liner may be offered for sale unless it is recyclable, biodegradable, photodegradable, or "otherwise degradable."

SOURCE: South Dakota Codified Laws: Title 5, Chapter 5-23 (State Purchases and Printing), §5-23-22.4 through 5-23-22.6; Title 7, Chapter 7-33 (Solid Waste Management Systems), §7-33-6; Title 13, Chapter 13-34, §13-34-26; Title 34A (Environmental Protection), Chapter 34A-6, §34A-6-1.2, §34A-6-1.5 and 1.6, Chapter 34A-7 (Litter Disposal and Control), §34A-7-5, §34A-7-5.1.

Tennessee

- The state's recycling goal was a 25 percent reduction of solid waste by 31 December 1995.
- The Division of Energy (DOE) is directed to promote research and development into a number of fields, including resource recycling. The DOE is also responsible for promoting the development of material recycling, handling, and management systems.
- All university establishments must develop annual source reduction, recycling, and waste management plans. Additionally, their curriculum sequences must assure adequate treatment of issues pertaining to recycling, source reduction, and waste management for undergraduate courses, continuing education programs, and training courses for local officials.
- A property tax exemption is offered on property owned by nonprofit organizations that recycle or dispose of waste products in a manner other than landfill disposal.
- The Department of Health is directed to develop a grant program for research and development in the field of solid waste disposal technologies, which include source separation and recycling.

- The general assembly has declared that source reduction and recycling should be practiced to the greatest extent possible to lessen the state's dependence on landfills. As of 1 January 1991, the State Planning Office was to establish a comprehensive solid waste management plan for the state. The integrated solid waste management hierarchy is defined as (1) source reduction; (2) recycling, mulching, and composting; (3) incineration; and (4) landfill disposal.
- The state hopes to educate all waste producers and handlers about reducing the amount of solid waste destined for landfill disposal as much as possible and to promote markets for recyclable materials.
- The state plan had to include educational programs and encourage governmental entities to recycle and buy recycled materials. The State Planning Office was also directed to assist in preparing and adopting regional solid waste management plans by the nine development districts within the state. Each of those districts must also produce a recycling plan that, among other things, provides for the disposal of household hazardous wastes and for citizen education programs. The state plan was also expected to establish goals for the reduction of solid wastes disposed of in landfills. Beginning 1 March 1994, each region must submit an annual report to the state, including information on its recycling programs.
- Within each municipal solid waste planning district, a Municipal Solid Waste Planning Advisory Committee was to be established.
- The State Planning Office has also been required to establish guidelines for source reduction, source separation, and collection of recyclable materials in all state agencies, including the school system. It must also establish and maintain a statewide solid waste management database.
- The following goals were established for the purchase of recycled paper and paper products as percentages of the total volume of paper purchased by the state: at least 10 percent in 1990, at least 25 percent in 1992, and at least 40 percent in 1994.
- The state requires that at least 40 percent of the recycled fiber in newsprint and newsprint products come from postconsumer newspaper waste.
- With the funds available from the state's Solid Waste Management Fund, an Office of Cooperative Marketing for

Recyclables is expected to be established by 1 July 1992. This office is charged with preparing and maintaining recycling directories for both buyers and sellers of recycled materials, creating a database to assist marketing efforts, and maintaining all information necessary to promote recycled materials markets within the state. Additionally, a Recycling Market Advisory Council will also be created to further assist in marketing efforts.

- The State Planning Office must establish an information clearinghouse on source reduction and recycling and conduct statewide and regional conferences and workshops on these subjects and on solid waste management.
- The Department of Transportation was directed to study the feasibility of using recycled plastic materials in guardrail posts or right-of-way fence posts. The results of that study were to be submitted to the state by 1 January 1991.
- Grant programs are available to fund the purchase of recycling equipment by nonprofit organizations.
- All educational institutions must provide and encourage education on solid waste, source reduction, and recycling. The State Planning Office has established an awards program to recognize outstanding accomplishments in this area.
- By 1 January 1996, all counties are required to provide at least one site for the collection of recyclables, and such facilities must file annual reports to the state regarding details of their recycling efforts. The University of Tennessee has been directed to provide technical assistance for the design and management of these recycling programs.
- All state agencies have to recycle paper, aluminum cans, glass bottles, newsprint, plastic bottles, mixed paper, and steel "tin" cans to the maximum extent possible. Additionally, these agencies must purchase recycled products or products made from recycled materials whenever possible.
- As of 1 January 1995, the following items are banned from landfills and incinerators and are required to be recycled: whole waste tires, lead-acid batteries, and used oil and automotive fluids.

SOURCE: Tennessee Code Annotated: Title 4, Chapter 3, Part 7, §4-3-709 and 710; Title 7, Chapter 21, Part 3, §7-21-301, Chapter 54, §7-54-101, §7-54-103, Chapter 58 (Resource Recovery and Solid Waste Disposal), §7-58-101, §7-58-110; Title 12, Chapter 3 (Public

Purchases), Part 5, §12-3-531; Title 49 (Education), Chapter 7, Part 1, §49-7-121, Part 2, §49-7-202, Chapter 9, Part 4, §49-9-406; Title 67, Chapter 5 (Property Taxes), Part 2, §67-5-208, Part 6, §67-5-604; Title 68 (Safety and Health), Chapter 1, Part 1, §68-1-104, Chapter 31 (Solid Waste Disposal), Part 1 (Solid Waste Disposal Act), §68-31-103, Part 4, §68-31-401, §68-31-405, Part 5, §68-31-501 through 503, Part 6, §68-31-602 through 607, Part 8, §68-31-802, §68-31-803, §68-31-812, §68-31-815, §68-31-825 through 827, §68-31-841 through 846, §68-31-848, §68-31-854, §68-31-861, §68-31-863 through 866, §68-31-871 and 872, Part 9 (Solid Waste Authority Act of 1991), §68-31-902.

Texas

- Texas funds its recycling programs through a Solid Waste Fee Revenue of 50 cents per ton of solid waste disposed.
- The Department of Health has been directed to establish and administer a Recycling and Waste Minimization Office to provide technical assistance to local governments on recycling and waste minimization. The department must also work with the Department of Commerce to develop recycled materials and compost markets.
- An Interagency Coordination Council is charged with responsibilities that include considering "the use of incentives to encourage waste minimization and reusing and recycling of waste, and the use of resource recovery facilities."
- The state's integrated solid waste management hierarchy is prioritized as follows: (1) waste minimization, (2) reuse and recycling, (3) waste-to-energy or safe incineration, and (4) land disposal.
- Local governments are entrusted with the control of solid waste collection and disposal. Regional or local solid waste management plans must include efforts to recycle or reuse waste as well as identify other opportunities for waste minimization, reuse, or recycling. Each plan must also include recommendations for improving the regional or local programs.
- Funding for recycling projects is available from the Technical Assistance Fund.
- Sludge recyclers are exempt from a franchise tax.
- The state requires that all products purchased by its agencies contain the highest amounts of recycled materials possible. A price preference of 15 percent is allowed in the purchase of highway paving materials made of recycled tires. Furthermore, all state agencies are expected to collect and recycle

their wastepaper and to participate in waste minimization programs.

- Plastic bottles and other rigid plastic containers sold in the state must carry a code identifying the resin they are made of (per Society of the Plastics Industry specifications). Violators of this law are subject to a civil penalty.
- The use of recycled paper in the court system is "strongly encouraged."

SOURCE: Vernon's Texas Statutes and Codes Annotated: Health and Safety Code, Title 5 (Sanitation and Environmental Quality), Subtitle B, Chapter 361, Subchapter A, Subchapter B, §361.014, §361.0151, §361.021 through 023, Subchapter C, §361.0861, Chapter 363 (Municipal Solid Waste), Subchapter A, §363.003, §363.006, Subchapter D, §363.064, Subchapter E, §363.095, Chapter 368, Subchapter B, §368.013; Tax Code, Title 2, Subtitle F (Franchise Tax), Chapter 171, Subchapter B, §171.085; Civil Statutes, Title 20, Article 1, §3.21, §3.211, Title 70, Chapter 8, Article 4413(33b), §2 and §3, Title 71, Chapter 4A, Article 4477-7G (Coding of Plastic Containers). Texas Rules of Civil Procedure, Part II, Section 4A, Rule 45. Texas Rules of Appellate Procedure, Section 2, Rule 4.

Utah

- As of April 1996, Utah's recycling rate was 13 percent.
- The Solid and Hazardous Waste Control Board was directed to establish a comprehensive statewide solid waste management plan by 1 July 1993. The plan was to include an assessment of the state's ability to minimize and recycle wastes.
- The state's chief procurement officer was to provide, on or before 1 June 1990, guidelines and requirements on the purchase of recycled paper and paper products for state agencies. Recycled paper is defined as any paper containing no less than 50 percent secondary wastepaper fibers (by weight). State agencies are given a 5 percent pricing preference for the purchase of recycled paper and paper products. Goals for purchasing specific percentages of the total amount of paper bought by the state per year begin with a 5 percent minimum requirement in fiscal year 1990–1991. The goal rises by 5 percent each year thereafter until the minimum purchase requirement is 50 percent.
- State agencies were also required to evaluate their potential for recycling paper by 1 October 1990. Subsequently, they were

to implement recycling programs if they could be run in a cost-effective manner.

- The legislature encourages waste tire recycling and the development of the recycling industry. As of 1 July 1990, a recycling fee was imposed on each new tire purchased from a retailer. The fee varies according to the size of the tire: 14 inches or less is $1, and greater than 14 inches is $1.50. The proceeds from this fee are deposited into the Waste Tire Recycling Expendable Trust Fund, which may be used to partially reimburse persons for the costs of using waste tires, waste chips, or similar products to create energy or a new product. This Waste Tire Recycling Act was repealed on 1 July 1995.
- As of 1 January 1992, lead-acid batteries could no longer be disposed of and had to be recycled. Battery sellers must post this law, act as collection centers, and accept up to two old batteries when a new one is purchased.
- The legislature encourages the recycling of used oil and the use of recycled oil.
- Utah has designed a Recycling Marker Development Zone program (H.B. 249), which includes provisions to encourage the growth of recycling business with the following incentives: business tax credits, simplified procedures for permitting, favorable zoning assistance, and possible utility rate concessions.

SOURCE: Utah Code: Title 19 (Environmental Quality Code), Chapter 6, Part 1 (Solid and Hazardous Waste Act), §19-6-102, §19-6-104, Part 5 (Solid Waste Management Act), §19-6-502 and 503, Part 6 (Lead Acid Battery Disposal), §19-6-602 through 605; Title 26 (Health Code), Chapter 32a (Waste Tire Recycling Act), §26-32a-101 through 113; Title 40, Chapter 9 (Oil Rerefinement Act), §40-9-2 through 3.5; Title 63, Chapter 55, Part 2 (Repeal Dates), §63-55-226, Chapter 56, Part B, §63-56-9, Part D, 63-56-20.7. H.B. 249 (Utah Recycling Market Development Zones).

Vermont

- As of April 1996, Vermont's recycling rate was 29 percent.
- The state has a recycling goal of 40 percent by 2000.
- The governor created a state recycling program to promote source reduction and the recovery of recyclable and reusable materials by state agencies in 1987. Documentation of com-

pliance was required by 15 January 1987 and must be submitted each odd-numbered year after that.

- At the same time, the director of purchasing was directed to review and modify all procurement specifications to encourage the purchase of products made from recycled materials. Recycled paper is defined as that containing not less than 50 percent postconsumer waste. Other recycled products include retreaded automobile tires, compost materials, and rerefined oils.
- The state's Solid Waste Management Plan must include goals that emphasize the greatest possible amount of source reduction, reuse, and recycling. Additionally, the state encourages educational programs, investment and employment opportunities, and industries and businesses that promote recycling and source reduction.
- The state must provide technical and financial leadership to municipalities regarding solid waste plans. Such plans must give priority to reducing the waste stream through recycling and to reducing nonbiodegradable and hazardous ingredients.
- A Solid Waste Management Advisory Committee was created in 1986 to assist the state in managing its solid waste.
- As of 1 July 1991, an integrated solid waste management plan was to be published, using the following hierarchy: (1) source reduction, (2) reuse and recycling, (3) waste processing to reduce the volume of waste, and (4) landfill disposal of residuals. This plan is to be revised at least once every five years.
- Vermont has a mandatory deposit-redemption law for all beverage containers.
- Funding for recycling and other solid waste disposal alternatives is available through the state secretary's office and through the Waste Management Assistance Fund, which is administered by the secretary of the Agency of Natural Resources.
- Any solid waste district that wishes to be eligible for grant monies must include a source separation plan in its solid waste management plan. A recycling awareness component for public education is also required.
- Additional incentives for source separation include the stipulation that unless ordinances mandating source separation are effective prior to 1 July 1993, the district or municipality is no longer eligible for grant funding. The sooner the ordinances are in effect, the more money the agency is eligible to receive.

- The secretary of natural resources has been informed that on 15 January 1991 and every year thereafter, a report must be submitted advising the legislature on the status of the newsprint industry's purchase of recycled newsprint. The Vermont Newspapers Users' Task Force set goals for the annual postconsumer fiber consumption of recycled newsprint: 11 percent by 1992 and 23 percent by 1995.
- Towns and cities are responsible for operating and maintaining their own solid waste facilities, including recycling centers. If a private recycling center existed prior to the municipality's, the government center may not adversely affect it.
- State purchasing agents are advised to purchase items made from recycled materials whenever possible. A 5 percent purchasing incentive is provided, and this may be exceeded if permission is obtained from the commissioner of general services. The goal for the state purchase of recycled materials is 40 percent by the end of 1993.
- Recycled materials accepted for recycling are exempt from the state's franchise tax on waste facilities.

SOURCE: Vermont Statutes Annotated: Title 3, Chapter 7, Executive Order No. 17 and 24; Title 10, Part 2, Chapter 53 (Beverage Containers; Deposit-Redemption System), Part 5, Chapter 159 (Waste Management), Subchapter 1, §6601, §6602, §6603b, §6603e, §6603f, §6604, §6605, §6618, §6619, §6622, §6622a; Title 24, Part 2, Chapter 61, Subchapter 8 (Rubbish and Garbage), §2203a and b, §2206; Title 29, Part 2, Chapter 49, §903; Title 32 (Taxation and Finance), Part 3, Chapter 151, Subchapter 13, §5953.

Virginia

- As of April 1996, Virginia's recycling rate was 33 percent.
- The state's recycling goal was 25 percent by 1995.
- Solid waste regions have been delineated, and each region is responsible for developing and implementing its own comprehensive solid waste management plan. The plans must include waste reduction, recycling, and reuse of solid wastes.
- The state's Litter Control and Recycling Program is to encourage recycling to the "maximum practical extent."
- As of 1 July 1992, every plastic bottle and rigid plastic container must carry a symbol identifying the resin it is made from (per the Society of the Plastics Industry specifications).

- On and after 1 January 1993, beverage containers that are connected with any plastic rings or devices that are not recyclable or degradable may not be sold.
- The Department of Litter Control and Recycling has been directed to develop and implement a plan to deal with waste tires.
- Lead-acid batteries may not be disposed of and must be recycled. Battery sellers must post signs that advise customers of this law and of the fact that their establishments are required to accept used motor vehicle batteries.
- All state agencies and universities had to establish recycling programs by 1 July 1991. The programs were to include the collection of used motor oil, glass, aluminum, office paper, and corrugated cardboard.
- All state agencies must buy recycled paper and paper products whenever possible. (Recycled paper is defined by Environmental Protection Agency standards.) A 10 percent price preference is permitted for state and local agencies alike.
- By 1 July 1992, the Department of Education must develop guidelines for source reduction and recycling programs in public schools, including the use of recycled materials.
- The Department of Economic Development has been directed to encourage and promote the recycling industry within the state.
- The Department of Transportation is encouraged to purchase re-refined or recycled motor oil. The recycling of used oil is promoted but not required.
- Counties, cities, and towns are permitted to require source separation and the collection of recyclables in both the residential and commercial sections of their jurisdictions.
- A 10 percent income tax credit was available for both businesses and individuals for the purchase of recycling equipment between 1 January 1991 and 1 January 1993. Qualified recycling equipment may also be exempt from property taxes.
- The Virginia Resources Authority was created in part to encourage the investment of both public and private funds in recycling and resource recovery projects.

SOURCE: Code of Virginia: Title 2.1, Chapter 32, Article 3 (Purchases and Supplies), §2.1-457; Title 10.1 (Conservation), Subtitle II, Chapter 14 (Virginia Waste Management Act), Article 1, Article 2, §10.1-1408.1, §10.1-1411, Article 3 (Litter Control and Recycling),

§10.1-1414 through 1425; Title 11 (Contracts), Chapter 7, Article 2, §11-47.2; Title 15.1 (Cities, Counties, and Towns), Chapter 1, Article 2, §15.1-11.5, §15.1-11.5:01, §15.1-11.5:2 and 3, §15.1-28.01; Title 33.1, Chapter 1, Article 15, §33.1-189.1; Title 45.1, Chapter 26, §45.1-390.1; Title 58.1 (Taxation), Subtitle I, Chapter 3, Article 3, §58.1-338, Article 14, §58.1-445.1, Subtitle III, Chapter 36, Article 5, §58.1-3661; Title 62.1, Chapter 21 (Virginia Resources Authority).

Washington

- As of April 1996, Washington's recycling rate was 38 percent.
- The state's recycling goal was 50 percent by 1995.
- According to state policy, used oil must be collected and recycled and public education programs to promote such recycling must be developed.
- Cities and towns are empowered to enact ordinances that deal with solid waste or collection of recyclable materials.
- All state and local agencies are encouraged to give preference to goods made from recycled materials in their purchasing activities.
- The Department of Trade and Economic Development is to assist in establishing and improving recyclable materials markets both inside and outside the state. The Washington Committee for Recycling Markets was created within this department in 1989.
- Washington's Model Litter Control and Recycling Act was passed in 1971 to promote litter control and to stimulate private recycling programs throughout the state. The act gives the Department of Ecology the authority to fulfill these goals. The state must also create a competitive awards program for source reduction and recycling programs in the public schools.
- Local governments are responsible for solid waste handling, but a comprehensive statewide plan, which includes recycling, is mandated as well. The state takes responsibility for providing technical and financial assistance to local governments for solid waste projects.
- Each county is responsible for preparing its own comprehensive solid waste management plan, and all cities within each county are directed to cooperate with such plans. Guidelines specifically dealing with waste reduction and recycling had to be submitted to the state by 15 March 1990 for approval.

Within these plans, each county could decide what materials should be recycled. In addition, recycling and trash collectors must set rates that encourage recycling and discourage waste disposal (e.g., volume-based rates), and drop-off or buy-back recycling centers must be provided in rural areas.

- The recycling of waste tires is encouraged. A $1 fee is levied on the sale of new tires, to be collected by retailers and ultimately deposited in the Vehicle Tire Recycling Account. Grants from this fund may be allotted to projects dealing with tire recycling.
- The Department of Ecology has been directed to provide educational materials that promote the reduction and recycling of household hazardous waste.
- Lead-acid batteries may not be disposed of or incinerated and must be recycled. Written notices to this effect must be posted at all battery-selling locations, and old batteries are to be accepted when new ones are bought on a one-to-one exchange basis. Additionally, if a person does not surrender a used battery when buying a new one, retailers are required to charge an extra $5 (core charge).
- Grant monies have been appropriated to study the composting of food and yard wastes. Yard waste collection programs must be provided where composting markets are available.
- A 1 percent state tax on garbage bills was exacted until June 1993. Funds derived from this tax were to be spent on technical assistance and grants to local governments to fund research on hard-to-recycle materials and market development.

SOURCE: West's Revised Code of Washington Annotated: Title 19, Chapter 19.114 (Used Automotive Oil Recycling); Title 35 (Cities and Towns), Chapter 35.21, §35.21.130, §35.21.158, Chapter 35.22, §35.22.620, Chapter 35.23, §35.23.352; Title 36 (Counties), Chapter 36.32, §36.32.250, Chapter 36.58 (Solid Waste Disposal), §36.58.010, §36.58.040, §36.58.160; Title 39, Chapter 39.30, §39.30.040; Title 43, Chapter 43.19, §43.19.1911, Chapter 43.31, §43.31.545, §43.31.552, Chapter 43.160, §43.160.077; Title 70 (Public Health and Safety), Chapter 70.93 (Model Litter Control and Recycling Act), Chapter 70.95 (Solid Waste Management—Recovery and Recycling), Chapter 70.95C (Waste Reduction), §70.95C.110, §70.95C.120; Title 81 (Transportation), Chapter 81.77 (Garbage and Refuse Collection Companies).

West Virginia

- As of April 1996, West Virginia's recycling rate was 13 percent.
- The state has a recycling goal of 30 percent by 2000.
- West Virginia's public policy includes provisions to establish comprehensive solid waste management plans with recycling as a component. To further these plans, a West Virginia Solid Waste Management Board was established in 1989.
- The responsibility for developing and implementing the solid waste management plan is assigned to the Division of Natural Resources (DNR). The DNR is also responsible for the West Virginia Litter Control Program, which encourages recycling. The legislature adds that "resource recovery and recycling reduces the need for landfills and extends their life and proper disposal, resource recovery, or recycling of solid waste is for the general welfare of the citizens of this state." The state's integrated solid waste management hierarchy is as follows: (1) source reduction; (2) recycling, reuse, and resource recovery; (3) environmentally acceptable incineration; and (4) landfill disposal.
- Each county and regional solid waste authority was responsible for developing and submitting to the state its own litter and solid waste control plan by 1 July 1991. Such plans had to include public education programs, timetables, and innovative incentives to encourage recycling efforts.
- Additionally, the West Virginia Recycling Act of 1989 requires each county to adopt a comprehensive recycling program for solid waste if the citizens vote to do so by petition and referendum. Such programs must encourage source separation, increase the purchasing of recycled products by local agencies, and educate the public about the benefits of recycling. Counties may decide which materials to recycle.
- A solid waste fee of $1.25 per ton funds solid waste and recycling programs. Any waste that is being reused or recycled is exempt from this fee.
- All state agencies must establish recycling programs for aluminum containers, glass, and paper, and any other recyclable materials that can be practicably collected and sold should be included in these programs. Such agencies are also required to encourage and promote recycling within all public facilities.
- State agencies are encouraged to purchase recycled products to the maximum extent possible.

- As of 1 June 1994, lead-acid batteries were banned from solid waste facilities.
- As of 1 June 1996, tires were banned from solid waste facilities.
- As of 1 January 1997, yard waste (including grass clippings and leaves) was banned from solid waste facilities.

SOURCE: West Virginia Code 1966: Chapter 16 (Public Health), Article 26 (West Virginia Solid Waste Management Board); Chapter 20 (Natural Resources), Article 5F (Solid Waste Management Act), Article 9 (County and Regional Solid Waste Authorities), Article 11 (West Virginia Recycling Program/Act).

Wisconsin

- As of April 1996, Wisconsin's recycling rate was 28 percent.
- The state has a recycling goal of 60 percent by 2000.
- Counties and municipalities were required to work together on solid waste management plans and advise the state by January 1993 as to how they will comply with the various landfill disposal bans and recycling requirements.
- Wisconsin expects to attain its solid waste disposal goals by banning numerous wastes from landfill disposal, including: lead-acid batteries, white goods, and waste oil by 1991; yard waste by 1993; and corrugated paper and other paperboards used for containers, aluminum and glass containers, polystyrene foam, magazines and other coated papers, newsprint, office paper, plastic and steel containers, and waste tires by 1995.
- State and local agencies must write their procurement specifications to promote the purchase of recyclable and recycled products. Additionally, single-use, disposable products are to be discouraged. Recycled content bid preferences of 1 percent for each 10 percent of recycled content in a product are permitted. By 1995, all recycled paper that is purchased must contain 40 percent recycled fiber.
- Newspaper publishers must use newsprint with at least 10 percent recycled content in 1991, progressing to 45 percent in 2001.
- Plastic containers must contain at least 10 percent recycled materials by 1993 and 25 percent by 31 January 1995.
- Cloth diapers and charges for diaper services are exempt from sales and use taxes.

- Beginning in the 1991–1992 fiscal year, a gross receipts tax was levied on businesses to fund recycling programs through a recycling fund.
- The Division of Natural Resources is to establish the state's priorities for developing markets for postconsumer waste materials.
- In 1996, the State Legislative Council formed a Special Committee on the Future of Recycling, to develop recommendations for legislative changes to Wisconsin's recycling law.

SOURCE: Wisconsin Statutes Annotated: Chapter 159 (Solid Waste), Subchapter II (Solid Waste Reduction, Recovery, and Recycling). Reindl, John, "Wisconsin's Recycling Law under Revision by Legislative Council," *AROW* (December 1996): 1–4.

Wyoming

- As of April 1996, Wyoming's recycling rate was 4 percent.
- At this time, the only recycling legislation in this state prohibits the disposal of lead-acid batteries and requires that they be recycled. Battery sellers must accept old batteries when new ones are purchased, and signage must be posted to this effect.

SOURCE: Wyoming Statutes 1977: Title 35 (Public Health and Safety), Chapter 11 (Environmental Quality), Article 5 (Solid Waste Management).

Additional Sources for Chapter 5

Heumann, Jenny M. "Will a Bottle Bill Pass in Georgia?" *Waste Age's Recycling Times* (20 January 1997): 14.

Raymond, Michele. "Associations Don't Expect Much Action in 1997." *State Recycling Laws Update* (December 1996). Internet: http://www.raymond.com/recycle.

———. "Cities Are Livid over PET Problems." *State Recycling Laws Update* (December 1996). Internet: http://www.raymond.com/recycle.

State, Federal, and Private Recycling Organizations

State Recycling Associations and State Government Agencies

Alabama

Alabama Department of Environmental Management (ADEM)
Land Division—Recycling Program
1751 Congressman W. L. Dickinson Drive
Montgomery, AL 36109-2608
Mailing address: P.O. Box 301463
Montgomery, AL 36130-1463
(334) 270-5651
(334) 279-3050 (FAX)
WWW: http://alaweb.asc.edu/govern.html
Michael W. Forster, State Recycling Coordinator

The mission of the Land Division of ADEM is "to ensure proper management of waste currently being generated, to ensure the correction of problems at sites where past improper waste management has resulted in threats to public health and/or the environment, and to coordinate planning for future waste disposal." ADEM's Recycling Program offers a number of recycling services to the public. These include educational presentations to both the general population and school students, recycling market assistance, and information on recycled product procurement. ADEM also runs a grant program for local government agencies and nonprofit organizations that are starting or expanding recycling programs.

PUBLICATIONS: Printed materials and fact sheets on general recycling, household hazardous waste disposal, and motor oil recycling, articles on composting yard wastes, and a recycling poster are available through ADEM.

Alabama Recycling Coalition (ARC)
P.O. Box 241373
Montgomery, AL 36124-1373
(205) 351-7760
(205) 849-4718
(205) 351-7764 (FAX)
Cindy Morehead, President

The ARC was founded in 1992 and is a statewide, nonprofit organization uniting the public and private sectors, as well as individuals, in the fields of waste reduction, material reuse, recycling, and composting. The ARC encourages networking, offers training and education, and works to heighten public awareness about solid waste issues.

PUBLICATIONS: The *ARC News*, a newsletter, is published every four months, and a membership directory is published each fall.

Alaska

Alaska Center for the Environment (ACE)
519 W. 8th Avenue, Suite 201
Anchorage, AK 99501
(907) 274-3621
(907) 274-8733 (FAX)
Holly Kane, Executive Director

The nonprofit ACE promotes "sound environmental policy, conservation of the state's natural resources, and education of the public and youth for wise decision-making." Encouragement of recycling and waste reduction is a part of these goals. ACE also coordinates the voluntary "Green Star" program that encourages schools and businesses to "incorporate techniques of waste reduction, energy conservation, and pollution prevention into their daily activities."

PUBLICATIONS: ACE offers tip sheets, including "Activities to Promote the 3R's" and "Green Star Daily Wake Up Call," in addition to a monthly newsletter, *The Future Is R's.*

Alaska Department of Environmental Conservation (ADEC)
Recycling Division
P.O. Box O
Juneau, AK 99811-1800
(907) 269-7582
(907) 269-7600 (FAX)
In-state hotline: 1-800-478-2864 (in Anchorage: 276-2964)
WWW: http://www.state.ak.us/
David Wigglesworth, Manager of Pollution Prevention

Alaska embarked on a formal pollution prevention campaign in 1990, by legislative initiative. At that time, ADEC was directed to "actively promote the waste management practices of source reduction and recycling." In response, ADEC created a Pollution Prevention Program that subsequently researched recycling and solid waste management in Alaska. The program's intent is to provide nonregulatory technical assistance.

PUBLICATIONS: In 1991, the program published a *Pollution Prevention Resource Guide.* The guide defines the program and lists a variety of other state agencies involved in recycling projects, as well as recycling equipment manufacturers and scrap dealers. Additionally, there are lists of recycling centers throughout the state and suppliers of recycled products.

Arizona

Arizona Department of Commerce (ADOC)
Phoenix City Square
3800 N. Central Avenue, Suite 1400
Phoenix, AZ 85012
(602) 280-1334
(602) 280-1338 (FAX)
Statewide toll-free hotline: 1-800-947-3873
Gregory R. Fisher, National Marketing Representative, Recycling

This department deals primarily with the markets development and business side of recycling, sponsoring programs ranging from

Recycling Equipment Tax Credits to Waste Reduction Grants. Through the ADOC, Arizona actively works toward attracting and encouraging recycling businesses.

PUBLICATIONS: The Department of Commerce offers a number of fact sheets and promotional materials.

Arizona Department of Environmental Quality (ADEQ)
Recycling Program, 5th Floor
3033 N. Central Avenue
Phoenix, AZ 85012
(602) 207-4865
(602) 207-4467 (FAX)
Recycling hotline: (602) 207-4463
Statewide toll-free hotline: 1-800-234-5677
Alisa Jellum, Recycling Unit Manager

The ADEQ was created in 1990 by the Arizona Solid Waste Recycling Act in order to respond to the state's growing solid waste management and recycling needs. Since its inception, the ADEQ has developed two waste reduction grant programs, started developing a statewide Household Hazardous Waste program, created a state recycling hotline, sponsored a composting workshop, and assisted recycling efforts in both the public and private sectors.

PUBLICATIONS: The ADEQ produces an annual report and provides access to numerous manuals, videotapes, and state documents dealing with Arizona's recycling interests. The *Recycling Review,* published jointly by Arizona State University and the Arizona Recycling Coalition, is also available through the ADEQ.

Arkansas

Arkansas Department of Pollution Control and Ecology (ADPC&E)
Solid Waste, Recycling, and Marketing Divisions
P.O. Box 8913
Little Rock, AR 72219-8913
(501) 682-0812
(501) 682-0880 (FAX)
Robert Hunter, Recycling Coordinator

Three divisions of the ADPC&E have programs relating to recycling and solid waste management. The Solid Waste Division regulates landfills and manages the solid waste planning and recycling grant program. The Recycling Division is charged with facilitating the development of a cooperative regional infrastructure for the collection, processing, quality control, and transportation of recyclables. Arkansas created 17 Regional Solid Waste Management Districts in 1991 that are overseen by the ADPC&E. The Marketing Division was established with a mission to develop markets within the state for postconsumer recyclables and to maintain an extensive database of markets and specifications.

PUBLICATIONS: The *Arkansas Waste Line,* the newsletter of solid waste, recycling, and market development, is published quarterly by the ADPC&E and the University of Arkansas Cooperative Extension Service. Specific, locally pertinent educational materials are distributed throughout the state, and the ADPC&E maintains a reference collection of materials on various aspects of recycling, including "pay-as-you-throw" programs. The ADPC&E has also published *Recycling and Source Reduction,* the *Recycling Reference Guide, A Manual for Planning and Implementing Community Recycling in Arkansas,* and a report of the Arkansas Solid Waste Fact Finding Task Force entitled *Solid Waste Strategies.*

Arkansas Recycling Coalition (ARC)
P.O. Box 25734
Little Rock, AR 72221-5734
(501) 227-6979
(501) 562-2541 (FAX)
E-mail: recyclearc@aristotle.net
Charles Cervants, President

The ARC is "dedicated to the long-term development of recycling and waste reduction in Arkansas" and is comprised of citizens from both the public and private sector, united to promote recycling education and programs. The ARC is affiliated with the National Recycling Coalition.

PUBLICATIONS: The ARC produces a quarterly newsletter, and provides an *Arkansas Buy Recycled Business Alliance Information Packet* upon request.

California

California Department of Conservation (CDC)
Division of Recycling
P.O. Box 944268
Sacramento, CA 94244-2680
(916) 323-3836
(916) 327-2144 (FAX)
In-state hotline: 1-800-RECYCLE
WWW: http://www.consrv.ca.gov/
Lawrence Goldzband, Director

The Division of Recycling oversees the state's recycling programs in both the public and private sectors and publishes resource papers on various aspects of recycling, as well as an annual report on the California Beverage Container Recycling and Litter Reduction Act.

PUBLICATIONS: The CDC's Division of Recycling produces a *Resource Center Guide* and the bimonthly technical bulletin, *California Recycling Review.* The division also maintains a library of books, periodicals, and videotapes for public use. A computer bulletin board service, *InfoCycle,* is another free service offered by the CDC.

California Integrated Waste Management Board (CIWMB)
8800 Cal Center Drive
Sacramento, CA 95826
(916) 255-2369
In-state recycling hotline: 1-800-553-2962
WWW: http://www.ciwmb.ca.gov
Daniel Pennington, Chairman

The CIWMB is responsible for overseeing California's waste management efforts, the highest priorities of which are "source reduction, recycling, and composting." The board is responsible for "developing strategies for reducing the amount of waste generated" in-state and for implementing solid waste legislation. CALMAX, a statewide materials exchange program, is also run by the board.

PUBLICATIONS: The board publishes a monthly newsletter, the *CIWMB Update,* and both publishes and distributes educational literature on solid waste solutions. A CALMAX catalog is published bimonthly.

Californians Against Waste Foundation (CAWF)
926 J Street, Suite 606
Sacramento, CA 95814
(916) 443-8317
(916) 443-3912 (FAX)
Mark Murray, Executive Director

CAWF is a nonprofit, tax-exempt organization that is "working to create a renewable, sustainable economy by promoting the reuse, recycling, and renewal of our natural resources." The foundation's goal is for California to reach a 50 percent recycling and waste reduction by 2000. It is working on every aspect of promoting and implementing recycling and source reduction programs that other states will use as models. CAWF's sister organization, Californians Against Waste (CAW), is a public interest citizens' lobby working to create "a recycling economy through the passage of state and local legislation." CAW led the successful fight to establish the state's bottle bill. CAW is California's only grassroots organization that lobbies full-time both locally and in Washington, D.C., for waste reduction and recycling policies.

PUBLICATIONS: CAWF's publications include: *How-To Guide for Establishing Curbside Recycling Programs, Guide to Recycled Printing and Office Paper, Shopper's Guide to Recycled Products, Buy Recycled: Keep a Good Thing Going,* and, *Do You Think Recycled Products Are. . . .*

Colorado

Colorado Office of Energy Conservation (COEC)
1675 Broadway, Suite 300
Denver, CO 80202-4613
(303) 620-4292; (303) 620-4288
WWW: http://www.state.co.us/gov_dir/energy_gov.html
In-state hotline: 1-800-632-6662
Kelly Blair Roberts, Recycling Coordinator

COEC coordinates and tries to facilitate the state's efforts in all areas of conservation, with a special division devoted solely to solid waste. COEC provides grants for a variety of educational projects and sponsors conferences on recycling on a regular basis (e.g., the "Colorado Buy Recycled Conference" and the "Paper Grades Workshop").

PUBLICATIONS: *Recycle Colorado* is a quarterly newsletter published by COEC. The agency also publishes and/or provides funding for additional literature on recycling, including *Buy Recycled: The Businessperson's Guide, Recycling Office Waste Paper—A Step-by-Step Guide*, and *Decision Maker's Guide: Waste Reduction and Recycling*.

Colorado Recycles (CR)
8745 W. 14th Avenue, #216
Lakewood, CO 80215-4850
(303) 231-9972
Dianne Beal, Executive Director

A nonprofit educational organization that promotes statewide recycling efforts, CR is dedicated to making Coloradans aware of why, where, and how they can recycle cans, bottles, newspapers, plastic containers, and other valuable materials. CR also provides educational material for schools and recycling information to Colorado residents.

PUBLICATIONS: CR produces a quarterly publication called *News to Re-Use* and publishes and distributes a recycling poster/guide to Colorado recyclers on an annual basis.

Connecticut

Connecticut Department of Environmental Protection (CDEP)
Bureau of Waste Management
79 Elm Street
Hartford, CT 06106-5127
(860) 424-3365
(860) 566-4924 (FAX)
Margaret "Meg" Enkler, Recycling Education Coordinator

CDEP's staff is committed to recycling assistance efforts. They are available to answer questions on state recycling policy, regionalization, interstate initiatives, grants administration, office paper recycling, recycling markets, commercial recycling, composting, collection systems and equipment, processing systems, enforcement, and public education.

PUBLICATIONS: Printed fact sheets, brochures, posters, children's activities, a traveling exhibit, and school assembly programs, all of which deal with recycling, are available from CDEP. Several videotape programs on recycling are also available for purchase. CDEP produces a monthly newsletter called the *Garbage Gazette*.

Connecticut Recyclers Coalition (CRC)
P.O. Box 4038
Old Lyme, CT 06371-4038
(203) 774-1253
(203) 779-2056 (FAX)
Winston Averill, President

CRC, a nonprofit organization, is composed of "a diverse group of individuals and organizations concerned with recycling and related issues in Connecticut." Created in 1989, CRC works to educate both its membership and the larger community about solid waste problems and the solutions of recycling, source reduction, composting, and market development. CRC also works to provide perspective on policies and programs to make recycling more effective, efficient, and economical. Six priorities have been identified in CRC's policy statement: market development; source reduction; organics; improving and enhancing existing programs; beyond the Mandate (to encourage recycling additional products beyond the scope of state law requirements); and household hazardous waste.

PUBLICATIONS: CRC publishes a quarterly newsletter, *Discards/ Discourse,* and makes a variety of educational materials available to the public.

Delaware

Delaware Economic Development Office (DEDO)
99 Kings Highway
P.O. Box 1401
Dover, DE 19903-1401
(302) 739-4271 x148
(302) 739-5749 (FAX)
Evadne Giannini, Recycling Economic Development Advocate

In 1995, DEDO and the Delaware Department of Natural Resources and Environmental Control (DNREC) combined forces to create the Green Industries Initiative program, which promotes the use of recycled materials and a reduction in waste generation within the state's manufacturing sector through the use of economic incentives such as income tax credits.

PUBLICATIONS: DEDO produces numerous Green Industries Initiative materials, as well as a state recycling directory and the *Paper Recycling Guide.*

Delaware Solid Waste Authority (DSWA)
1128 S. Bradford Street
P.O. Box 455
Dover, DE 19903-0455
(302) 739-5361
(302) 739-4287
In-state hotline: 1-800-404-7080
Rich Von Stetten, Manager of Recycling

DSWA operates the program Recycle Delaware, which operates a state-wide voluntary drop-off, source-separated recycling program. DSWA started this program in 1990 with the opening of its first recycling center; as of 1995, there were 120 such centers in operation.

PUBLICATIONS: DSWA produces an annual report for Recycle Delaware and a recycling pocket guide.

District of Columbia

District of Columbia Solid Waste Management Administration
Department of Public Works (DPW)
2750 S. Capitol Street, SE
Washington, DC 20032
(202) 645-7044
(202) 404-1311 (FAX)
Leslie A. Hotaling, Administrator

This department is in the process of restructuring and reprioritizing its recycling programs, due to lack of funding.

Florida

Florida Department of Environmental Protection (FDEP)
Division of Waste Management
Bureau of Solid and Hazardous Waste
Solid Waste Section
2600 Blair Stone Road, MS 4565
Tallahassee, FL 32399-2400
(904) 488-0300
(904) 921-8061 (FAX)
WWW: http://www.dep.state.fl.us/index.html
Mary Jean Yoen, Manager of Solid Waste Section

FDEP was created in 1993 with the merger of the Department of Natural Resources and the Department of Environmental Regulation for the purpose of protecting and conserving the state's environment and natural resources. The Solid Waste Section manages an extensive Pollution Prevention Program (PPP), which includes recycling education under its auspices. The PPP is nonregulatory and can provide educational resources, on-site technical assistance, speakers for workshops, and networking information.

PUBLICATIONS: Publications, fact sheets, and videotapes are available from the Pollution Prevention Program.

RecycleFlorida Today, Inc. (RFT)
1015 US 301 S., #2425
Tampa, FL 33619
(813) 441-6245
(813) 626-5865
WWW: http://www.enviroworld.com/Resources/RFT.html
E-mail: recyclefl@aol.com
Susan Lancianse, Chairperson

RFT is a nonprofit organization originally formed in 1991 as a state chapter of the National Recycling Coalition. RFT acts as a liaison between government, industry, and others involved in recycling. It is both a professional and educational organization and will assist individual members in improving skills and techniques in recycling.

PUBLICATIONS: RFT publishes a quarterly newsletter, the *Renewable News,* and provides its membership with the NRC's monthly newsletter, *The NRC Connection.*

Georgia

Georgia Department of Community Affairs (GDCA)
60 Executive Park South, NE
Atlanta, GA 30329-2231
(404) 679-4940
Paula R. Longo, Recycling Coordinator

Both the Environmental Protection Division of the Department of Natural Resources (below) and GDCA are charged by the General Assembly of Georgia with solid waste management responsibilities. GDCA serves as the lead state agency for municipal solid waste recycling, waste reduction, and public education efforts. This department has also implemented, and oversees, the Georgia Clean and Beautiful Program, as well as additional training programs for local recycling coordinators.

PUBLICATIONS: GDCA publications include the pamphlet *Home Composting in Georgia*, as well as fact sheets, a state *Recycling Markets Directory*, and an annual *Solid Waste Management Report*.

Georgia Department of Natural Resources
Environmental Protection Division (EPD)
Floyd Towers East
205 Butler Street, SE
Atlanta, GA 30334
(404) 362-2692
(404) 362-2654 (FAX)
WWW: http://www.state.ga.us/
Pam Thomas, Environmental Specialist in Charge of Recycling

EPD, a division of the Georgia Department of Natural Resources, is responsible for solid waste management with regard to landfill regulation, numerous regulatory duties, and "encouraging, as appropriate, the reduction, reuse, and recycling of solid waste." EPD deals directly with local governments and is in charge of coordinating the state's solid waste activities.

PUBLICATIONS: Georgia's publications on recycling are published by the Georgia Department of Community Affairs (above).

Georgia Recycling Coalition (GRC)
2508 Kiner Court
Lawrenceville, GA 30243
(770) 822-9308
(770) 338-2735
Stephanie Hubbard, Administrative Coordinator

GRC is a nonprofit organization whose goals include: "To support the implementation and continuation of sustainable reduction and recycling programs throughout the state; to promote reduction and recycling education and information exchange through its speaker's bureau, newsletter, and meetings; and to work with other organizations in the state to foster common goals through the establishment of a Recycling Coordinating Council."

PUBLICATIONS: GRC produces a quarterly newsletter entitled, *Georgia Recycles*.

Hawaii

Clean Hawaii Center (CHC)
c/o Department of Business, Economic Development, & Tourism
236 S. Beretania Street, Room 506
Honolulu, HI 96804
(808) 587-3811
(808) 587-3820 (FAX)
Carilyn Shon, Director

CHC is designed to promote recycling and remanufacturing markets through the development of local end-use businesses, cooperative marketing, cooperative purchasing of recycled products, and promotion of Buy Recycled in Hawaii. The center works with other state agencies, businesses, and nonprofit groups to divert recyclable materials from landfills and incinerators, and it supports the development of local businesses that manufacture products from wastepaper, glass, plastics, green waste, and construction and demolition materials. CHC is administered by the Energy, Resources, and Technology Division of the Department of Business, Economic Development, & Tourism, and provides technical and other support assistance, sponsors workshops, and assists in recycled product development.

PUBLICATIONS: CHC develops printed materials to assist industry development in the state; a list of publications is available on request.

Hawaii Department of Health
Environmental Management Division
Office of Solid Waste Management (OSWM)
919 Ala-Moana Boulevard, Room 210
Honolulu, HI 96814
(808) 586-4240
(808) 586-7509 (FAX)
WWW: http://www.hawaii.gov/health
John Harder, Solid Waste Coordinator

OSWM is directed to permit and monitor solid waste management facilities and to encourage and promote alternative waste management. It provides technical assistance to permitted facilities and county agencies, as well as initiating or implementing legislation that promotes the development of recycling. OSWM also has a public education and outreach program, and hosts or supports workshops that promote and encourage composting, reuse, waste minimization, and recycling. The office distributes funding to counties to support glass recovery and "do-it-yourself" oil recycling, and also provides financial and in-kind support to the Clean Hawaii Center (above) to support the economic development of recycling businesses and services.

PUBLICATIONS: Copies of the state's legislative acts regarding solid waste are available through OSWM, as is a list of recycling coordinators from each of the islands that compose the state.

Idaho

Association of Idaho Recyclers (AIR)
1020 Denver Street
Idaho Falls, ID 83402
(208) 529-8084
(208) 520-0964 (FAX)
Cristy Hamilton Eames, President

Founded in 1995, AIR's mission is "to facilitate the expansion, stability, and diversity of source reduction and recycling in the State of Idaho." AIR holds an annual state recycling conference, and workshops and seminars are offered on a regular basis.

PUBLICATIONS: AIR produces a quarterly newsletter, *What's Up in the AIR.*

North Central District Health Department
215 10th Street
Lewiston, ID 83501
(208) 799-0353
(208) 799-0349 (FAX)
Toll-free recycling hotline: 1-800-373-2099
WWW: http://www.idwr.state.id.us/oneplan/waste/handle.htm
Bill Lillibridge, Waste Reduction Coordinator

There are seven health districts in Idaho. The North Central District coordinates and has regulatory power over all solid waste activities throughout the state. This district is also responsible for promoting and encouraging source reduction and recycling programs.

Illinois

Illinois Bureau of Energy and Waste Management
Department of Commerce and Community Affairs (DCCA)
Division of Recycling
325 W. Adams Street, Room 300
Springfield, IL 62704-1892
(217) 785-2800
(217) 785-2618 (FAX)
Reg Willis, Manager of Program Management

The Illinois Solid Waste Management Act authorizes DCCA to "promote alternative means of managing solid waste with the goal of reducing reliance on land disposal." Through its Bureau of Energy and Waste Management, DCCA has established numerous grant programs for projects specifically aimed at its solid waste handling goals. Two of the grant programs concern themselves solely with recycling, while one deals with the disposal problems of used tires and another revolves around public education programs that teach about waste reduction and recycling.

PUBLICATIONS: A collection of waste reduction/recycling "clip art" graphics is found in a novel and useful booklet available from the Division of Recycling, which also maintains an information clearinghouse that publishes an extensive list of print and video resources on recycling, composting, and solid waste management. Publications are free, and videos are loaned to businesses and local governments only. However, individuals may borrow the videos through local libraries. The clearinghouse hotline, for in-state use, is 1-800-252-8955. The Division of Recycling also publishes two quarterly newsletters: *Recycling Update* and *The Three R's* (for elementary teachers).

Illinois Recycling Association (IRA)
9400 Bormet Drive, Suite 5
Mokena, IL 60448
(708) 479-3800
(708) 479-4592 (FAX)
E-mail: ILRecycle@aol.com
Nancy Burhop, President

IRA is a statewide coalition of individuals and organizations that provides for the exchange of information and ideas on resource conservation through recycling and waste reduction. As a state affiliate of the National Recycling Coalition, IRA sponsors workshops and conferences on recycling.

PUBLICATIONS: IRA's newsletter, *Material Matters*, is published ten months of the year and includes a section on regional markets.

Indiana

Indiana Department of Environmental Management (IDEM)
Department of Pollution Prevention and Technical Assistance
100 N. Senate Street
P.O. Box 6015
Indianapolis, IN 46206
(317) 232-8172
(317) 233-5627 (FAX)
Statewide toll-free hotline: 1-800-451-6027
Bob Gedert, Branch Chief of Source Reduction and Recycling

IDEM was established 1 April 1986 to protect the environment. IDEM's responsibilities include enforcing the Resource Conservation and Recovery Act, and its goals include "working with local solid waste management districts to reach the state goal of reducing solid waste 35 percent by 1996, and 50 percent by the year 2001." Indiana did not reach its 1996 goal of a 35 percent reduction in solid waste, but its goal for 2001 remains the same.

PUBLICATIONS: IDEM's Office of External Affairs produces numerous pamphlets for the public, including *Recycle Indiana, Cool Things Kids Can Do to Help Save the Earth, Simple Substitutes for Household Hazardous Products,* and *Composting Is Nature's Recycling System.*

Indiana Recycling Coalition (IRC)
P.O. Box 20444
Indianapolis, IN 46220-0444
(317) 283-6226 (phone and FAX)
WWW: http://www.papertrail.com/irc/
E-mail: recycling@in.net
Larry Wilson, President

IRC, a nonprofit corporation founded in 1989 that represents concerned citizens, local officials, business, industry, and environmental groups, is working to "close the recycling loop." IRC heads the campaign, "Indiana's Had It Up to Here," to deal with the state's solid waste issues and supports efforts to develop markets for products made from recyclables. In addition to organizing in-state recycling conferences, the coalition serves as an information clearinghouse and coordinates the efforts of those involved in the recycling cycle. IRC also promotes the use of recyclable and recycled materials, with a special emphasis on community education.

PUBLICATIONS: IRC publishes a monthly newsletter and *The Indiana Recycling Handbook* (containing everything you've always wanted to know about recycling in Indiana), which is updated every spring. Newsletter information is available both in print and by e-mail. Other publications include *Eat, Drink, and Recycle . . . A Guide to Recycling for Restaurants, Bars, and Clubs,* and *Reduce and Recycle, Indiana . . . Changing Our Habit of Waste.*

Iowa

Iowa Department of Natural Resources
Waste Management Authority Division (WMAD)
Wallace Building
900 E. Grand Avenue
Des Moines, IA 50319-0034
(515) 281-8941
(515) 281-8895 (FAX)
In-state recycling hotline: 1-800-532-1114
WWW: http://www.igsb.uiowa.edu/dnr_www.htm
Tom Blewett, Bureau Chief, Waste Reduction Bureau

WMAD was created in 1987 to assist Iowa in dealing with its waste management problems and to offer solutions. Its primary functions are to plan the state's long-term waste management needs; research and demonstrate waste management options and markets for recycling; implement new waste management programs; and provide public education, outreach, and technical assistance. WMAD works in a nonregulatory capacity with businesses, governments, and individuals to help them cope with waste-related problems, protect the environment, and prepare for the future. The division has extensive programs dealing with household hazardous waste, hazardous waste reduction, and solid waste recycling and reduction. (The state subscribes to the EPA hierarchy of integrated waste management.)

PUBLICATIONS: WMAD publishes a number of brochures and technical documents on recycling, composting, waste reduction, and household hazardous waste. It also produces videos on solid waste alternatives, pollution prevention, recycling, and composting. Videos geared toward specific age groups are also available for schools. Several state recycling directories are available through WMAD. A curriculum package for kindergarten through twelfth-grade classes, *Iowa Clean SWEEP,* prepared in cooperation with the Department of Education, is also produced by WMAD. The division works with the Iowa Waste Reduction Center (IWRC) at the University of Northern Iowa, which has a hotline to provide waste reduction information to businesses and industry: 1-800-422-3109 (see below).

Iowa Recycling Association (IRA)
200 E. Grand Avenue
Des Moines, IA 50309
(515) 242-4755
(515) 242-4749 (FAX)
E-mail: munderwo@ided.state.ia.us
Margo Underwood, President

This nonprofit organization, self-described as "newly evolving," is composed of members from both the public and private sectors. The state chapter of the National Recycling Coalition, IRA sponsors statewide conferences and bimonthly forums on recycling, in addition to monthly membership meetings. It has also cosponsored workshops and public education campaigns dealing with recycling.

PUBLICATIONS: IRA publishes a quarterly newsletter, the *Iowa Recycling Association Newsletter.*

Iowa Waste Reduction Center (IWRC)
University of Northern Iowa
75 Biology Research Center
Cedar Falls, IA 50614-0185
(319) 273-2079
(319) 273-2926 (FAX)
National toll-free information line: 1-800-422-3109
Jennifer Drenner, Program Manager

The IWRC, located at the University of Northern Iowa, provides free, nonregulatory technical assistance to small businesses in the areas of waste reduction and pollution prevention. The center also conducts applied research and education "that seeks practical solutions to small business environmental issues." Programs created and run by the center include the Small Business Pollution Prevention Center, the Mobile Outreach for Pollution Prevention, and the By-Product and Waste Search Service.

PUBLICATIONS: The IWRC produces a variety of publications on pollution prevention, most of which are geared toward small businesses. A list of these materials is available by calling 1-800-422-3109. The center also produces two quarterly newsletters: *Point Source,* which deals with hazardous wastes issues, and *The Closed Loop,* which addresses the reduction and/or recycling of numerous nonhazarous wastes.

Kansas

Kansas Department of Health and Environment (KDHE)
Division of Environment
Forbes Field, Building 740
Topeka, KS 66620-0001
(913) 296-1617
(913) 296-8909 (FAX)
Kent Foerster, Environmental Scientist

As of March 1997, the Division of Environment began its "Kansas—Don't Spoil It" campaign to educate the public on environmental issues, including responsible waste management. Composting is becoming an increasingly important issue in this state's waste management hierarchy, and as a result, more conferences and workshops have been developed to educate the state's citizens. A recycling grant program is administered by the KDHE as part of a larger, more inclusive Solid Waste Management Grants program.

PUBLICATIONS: Numerous posters, pamphlets, and fact sheets are being published for distribution with the "Don't Spoil It" program. A newsletter is also being designed.

Kentucky

Kentucky Department for Environmental Protection (KDEP)
Division of Waste Management
Resource Conservation and Local Assistance Branch
Frankfort Office Park
14 Reilly Road
Frankfort, KY 40601-1190
(502) 564-6716
(502) 564-4049 (FAX)
Joy Morgan, Manager

The Division of Waste Management wrote and submitted to the legislature the State Solid Waste Reduction and Management Plan in March 1992. The division's work for the next few years "will focus on enforcing and enhancing the regulations and statutes [on solid waste management] already in place, educating and training regulators and regulated, and cooperating with other state agencies who are pursuing similar goals." Currently, the division provides technical and informational assistance on recycling, and it networks with other state agencies. It also runs both a Clean Community Program (a Keep America Beautiful project) and a used oil recovery program. In addition, it is revising a solid waste curriculum guide for high schools and has a mobile "Eco-Home" that travels around the state for use by civic groups and schools. Additionally, the division administers the Kentucky Recycling and Marketing Assistance Program (KRMA), founded in 1995.

PUBLICATIONS: A periodic bulletin, *Waste Action,* is published by the division to provide information on the 3Rs. Another bulletin, *The Planning Edge,* deals with solid waste planning issues. A number of booklets, pamphlets, and information sheets are available that specifically deal with the KRMA program.

Kentucky Recycling Association (KRA)
2207 Eastern Avenue
Covington, KY 41014
(606) 356-8555
Anthony Noll, President

KRA is a nonprofit organization dedicated to promoting recycling throughout the state. It is composed of members from industry, private business, state and local government, and public service groups as well as interested citizens who seek to improve recycling and waste management practices within Kentucky. KRA itself is a member of the National Recycling Coalition.

PUBLICATIONS: *Recycling in State Government* is available from KRA, which also provides information on recycling and composting programs as well as a market directory and a county recycling directory. A newsletter, *The KRA Connection*, is published on a quarterly basis.

Louisiana

Louisiana Department of Environmental Quality (LDEQ)
Solid Waste Division
P.O. Box 82178
Baton Rouge, LA 70884-2178
(504) 765-0249
(504) 765-0299 (FAX)
WWW: http://www.deq.state.la.us/oshw/oshw.htm
Karen Fisher Brasher, Program Manager for Recycling and Waste Minimization

LDEQ is responsible for maintaining a "healthful and safe environment for the people of Louisiana." Within the Office of Solid and Hazardous Waste is the Solid Waste Division, which oversees a number of recycling programs dealing with a diversity of products, e.g., household paint, office papers, used oil, holiday cards, tires, and Christmas trees.

PUBLICATIONS: LDEQ publishes a directory of state recyclers and distributes a variety of EPA publications on solid waste and recycling.

Maine

Maine Resource Recovery Association (MRRA)
60 Community Drive
Augusta, ME 04330-9486
(207) 623-8428
(207) 626-5947 (FAX)
In-state toll-free hotline: 1-800-452-8786
Cathy Callahan, Executive Director

Founded in 1984, MRRA is an Affiliate of the Maine Municipal Association. The purpose of MRRA is to "promote professional solid waste management standards and practices and to further the development of recycling and other forms of resource recovery." MRRA provides information, a forum for the exchange of information, and technical assistance to its members. It also promotes market development and provides cooperative marketing opportunities. The John M. Howells Recycling Education Fund has been established by MRRA to provide funds for innovative recycling education programs.

PUBLICATIONS: MRRA produces a periodic newsletter, *Mainely Resources.*

Maine State Planning Office
Waste Management and Recycling Program
38 State House Station
Augusta, ME 04333-0038
(207) 287-3261
(207) 287-8059 (FAX)
WWW: http://www.state.me.us/spo
Kirk Goddard, Acting Executive Director

After six productive years, the state legislature closed the Maine Waste Management Agency in 1995 and transferred some of its functions and responsibilities to the State Planning Office. Specifically, the office must provide municipal technical assistance, data collection and management, grant administration, business assistance, and regular evaluation of Maine's various waste handling and reduction programs.

PUBLICATIONS: *Maine Waste Management: 20 Years of Progress* was published in 1995 as an overview of the state's accomplishments and future goals in solid waste management.

Maryland

Maryland Department of the Environment (MDE)
Recycling Services Division (RSD)
2500 Broening Highway
Building 40, 2nd Floor
Baltimore, MD 21224
(410) 631-3315
(410) 631-3321 (FAX)
In-state recycling hotline: 1-800-IRECYCLE
WWW: http://www.mde.state.md.us
Lori A. Scozzafava, Chief, Recycling Services Division

RSD, authorized by the Maryland Recycling Act of 1988, is expected to "reduce the disposal of material resources in Maryland through management, education, and regulation." Responsibilities of RSD include assisting county and state agencies with recycling; following local, national, and international markets for recyclable materials; managing and regulating scrap tires (as well as promoting tire recycling); tracking the state's recycling activities; and "fostering increased recycling through public education and policy development."

PUBLICATIONS: MDE publishes the pamphlet, *Reduce, Reuse, Recycle,* as well as several fact sheets on composting, the state's waste tire program, household hazardous waste, and recycling.

Maryland Recyclers Coalition (MRC)
584 Bellerive Drive, Suite 3D
Annapolis, MD 21401
(410) 974-4472
(410) 757-3809 (FAX)
WWW: http://www.marylandrecyclers.org
E-mail: mrc@mdassn.com
Marie Anderson, Executive Director

MRC is a nonprofit organization representing a diverse membership of public and private interests, as well as concerned citizens, who believe that "recycling is a viable and integral part of solid waste management." MRC is also a member group of the National Recycling Coalition. MRC provides education, outreach, an annual conference, and training programs to its constituents.

PUBLICATIONS: The *MRC Newsletter* is published on a bimonthly basis, and copies of "white papers" and policy documents prepared by the MRC board of directors are available on request.

Massachusetts

Massachusetts Department of Environmental Protection
Division of Solid Waste Management
1 Winter Street
4th Floor
Boston, MA 02108
(617) 292-5962
(617) 556-1049 (FAX)
WWW: http://www.state.ma.us/dep/bwp/dswm/dswmhome.htm
Robin Ingenthron, Recycling Programs Director

Massachusetts's Department of Environmental Protection (DEP) is the commonwealth's lead agency for solid waste management. Working with other offices within the Executive Office of Environmental Affairs and through the Division of Solid Waste Management, DEP's broad mission is to ensure that the state's solid waste system integrates source reduction, recycling, combustion with energy recovery, and environmentally safe operation of landfills. Recycling is promoted by the division through administration of an equipment grants program, as well as programs dealing with state government procurement of recycled products, the development of new markets for recyclables, and researching of available technologies for the handling of difficult-to-manage wastes. The division also offers technical assistance and outreach from staff on recycling and composting issues, and it acts as an information resource.

Massachusetts Recycling Coalition/MassRecycle
25 West Street, 2nd Floor
Boston, MA 02111
(617) 338-0244
(617) 338-8611 (FAX)
E-mail: massrecy@aol.com
Dorothy Suput, Executive Director

MassRecycle is a "statewide nonprofit coalition of individuals, local and state government, industry, and environmental groups dedicated to promoting and facilitating waste reduction, reuse, recycling, and buying recycled in order to achieve the environmental, social, and economic benefits derived from these activities." MassRecycle also organizes conferences and workshops, provides educational information, coordinates regular group tours of state-of-the-art recycling facilities, and loans a variety of resource materials, including videotapes (send for a current list). The Recycling Education Assistance to Public Schools (REAPS) program is administered by MassRecycle and provides recycling and solid waste education to teachers, students, and municipal recycling coordinators across the state.

PUBLICATIONS: MassRecycle publishes its newsletter, the *MassRecycler,* five times a year. A statewide guide to municipal recycling and composting is also available, as is *The Workplace Waste Reduction Guide.*

Michigan

Michigan Department of Environmental Quality (MDEQ)
Environmental Assistance Division (EAD)
P.O. Box 30457
Lansing, MI 48909-7957
(517) 373-1322
(517) 335-4729 (FAX)
Toll-free recycling hotline: 1-800-662-9278
WWW: http://www.deq.state.mi.us/
E-mail: doroshkl@deq.state.mi.us
Lucy Doroshko, Recycling Coordinator/Environmental Quality Analyst

Michigan's EAD provides assistance to state businesses, municipalities, institutions, and the general public, specifically in the areas of technical compliance, pollution prevention, waste reduction, and other environmental quality issues. Several education and outreach programs are administered by EAD, and its toll-free hotline (above) offers nationwide referral assistance on all MDEQ programs, as well as information on recycling and solid waste.

PUBLICATIONS: The MDEQ publishes a list of waste reduction and pollution prevention publications that are free to the public. This list can be ordered through the EAD hotline.

Michigan Recycling Coalition (MRC)
P.O. Box 10240
Lansing, MI 48901-0240
(517) 371-7073
(517) 371-1509 FAX
Kerrin O'Brien, Executive Director

MRC is a nonprofit, statewide organization dedicated to promoting sound solid waste management through recycling and reduction of waste material. It also promotes cooperation between the public and private sectors, encourages information exchange and technical assistance among recyclers, and works on public education, legislation, and marketing with respect to recycling. MRC also sponsors an annual recycling conference as well as workshops throughout the year.

PUBLICATIONS: MRC publishes a quarterly newsletter.

Minnesota

Minnesota Pollution Control Agency (MPCA)
520 Lafayette Road North
St. Paul, MN 55155-4194
(612) 282-2663
(612) 296-9707 (FAX)
Toll-free hotline: 1-800-657-3864
WWW: http://www.pca.state.mn.us
Bill Dunn, Policy Analyst

MPCA was created by the state legislature nearly 30 years ago to protect Minnesota's lakes and rivers. Since then, MPCA's authority has been expanded to include the protection of the air, groundwater, land, and human health. The mission of MPCA is "to protect Minnesota's environment to secure the quality of life of its citizens." MPCA works with many partners in communities, government, and industry to examine the quality of the state's environment, and helps the public and private sectors comply with state and federally mandated environmental laws.

PUBLICATIONS: MPCA publishes a series of fact sheets on solid waste subjects, as well as a bimonthly newsletter entitled *Minnesota Environment.*

Recycling Association of Minnesota (RAM)
P.O. Box 16467
St. Paul, MN 55116-0467
(612) 699-4004
(612) 422-0286 (FAX)
Donna M. Gray, Executive Director

RAM is a nonprofit organization representing a variety of individuals and other organizations that are "committed to the common goal of maximizing recycling as an integral part of waste and resource management." RAM provides technical assistance and educational programs to the state, and participates in research to further waste reduction goals in Minnesota through reuse, recycling, and composting. It is also a member organization in the National Recycling Coalition.

PUBLICATIONS: RAM produces a newsletter, *The Loop,* four to six times a year.

Mississippi

Mississippi Department of Environmental Quality (MDEQ)
Office of Pollution Control (OPC)
P.O. Box 10385
Jackson, MS 39289-0385
(601) 961-5171
(601) 354-6612 (FAX)
Larry Estes, State Recycling Coordinator

MDEQ/OPC organizes conferences and workshops on solid waste management, including recycling issues, and works with both the public and private sectors by providing educational information and a speakers bureau. The agency also funds the Mississippi Solid Waste Reduction Assistance Program (MSSWRAP) through Mississippi State University, offering technical assistance to citizens in the areas of waste minimization and reduction (telephone: [612] 325-8454).

PUBLICATIONS: MSSWRAP publishes a state recycling list and a monthly newsletter, the *MISSTAP NEWS;* it also runs a solid waste clearinghouse/library.

Mississippi Recycling Coalition (MRC)
c/o Keep Mississippi Beautiful
4785 I-55 North, Suite 103
Jackson, MS 39206
(601) 981-5566
1-800-545-3764
Barbara Door, Executive Director

MRC is a brand-new, nonprofit organization in the process of being founded in 1997. Its mission statement directs the organization "to complement, promote, and encourage responsible solid waste management, which includes recycling and composting programs, and to coordinate and unite the activities of professionals, organizations, business and industry, institutions, government agencies, and individuals concerned about solid waste management."

Missouri

Missouri Department of Environmental Quality (MDEQ)
Solid Waste Management Program
205 Jefferson
Jefferson City, MO 65102
(573) 751-5401
(573) 526-3902 (FAX)
Toll-free hotline: 1-800-334-6946
L. Kathleen Weinsaft, Chief of Solid Waste Planning Unit
Alice W. Geller, Market Development Program Director, EIERA

MDEQ oversees waste management programs throughout Missouri and works with the Environmental Improvement and Energy Resources Authority (EIERA) to promote recycling. EIERA is an independent agency created by the Missouri legislature "to provide low- or no-cost financing for projects that would reduce, control, and prevent environmental pollution, and to encourage research and development of energy alternatives while promoting economic development."

PUBLICATIONS: MDEQ publishes a quarterly magazine, *Missouri Resource Review.* A pamphlet on the recycling market development program is also available from MDEQ.

Missouri State Recycling Association (MORA)
101 West Tenth Street
Rolla, MO 65401
(573) 364-2993
(573) 364-7235 (FAX)
Lee Fox, President

MORA is a newly organized, nonprofit group that promotes waste reduction and recycling in the state by "providing information, educational opportunities and technical assistance in partnerships with national, state and regional organizations." MORA's goals include assisting consumers, businesses, and government in reducing their waste generation and increasing their recycling abilities, promoting recycled products and source reduction, developing local "waste-based" industries to manufacture products from recovered materials, promoting pollution prevention, and assisting in information exchange among state recyclers. MORA is a member of the National Recycling Coalition (NRC) and expects to have a World Wide Web site and e-mail address working as of September 1997.

PUBLICATIONS: MORA provides educational materials to its membership and publishes a quarterly newsletter, the *Missouri Recycling Association Newsletter.*

Montana

Associated Recyclers of Montana (ARM)
458 Charles Street
Billings, MT 59101
(406) 252-5721
(406) 252-8059 (FAX)
Mark Richlen, President

ARM is a nonprofit organization that works through the Keep Montana Clean and Beautiful program. It is composed primarily of recyclers in the beverage and solid waste industries.

PUBLICATIONS: ARM publishes a directory to the state's recycling centers.

Montana Department of Environmental Quality (MDEQ)
Planning, Prevention, and Assistance Division
Pollution Prevention Bureau
Metcalf Building
1520 E. 6th Avenue
(P.O. Box 200901)
Helena, MT 59620-0901
(406) 444-6697
(406) 444-6836 (FAX)
E-mail: pnelson@mont.gov
Peggy Nelson, Waste Reduction and Recycling Coordinator

The Pollution Prevention Bureau of the MDEQ is devoted to reducing or preventing pollution and managing wastes. This ties into the department's mission to "protect, sustain, and improve a clean and healthful environment to benefit present and future generations." Most of the bureau's activities are derived from state and national laws, and most services are federally funded as part of ongoing environmental programs. The bureau's projects are planned and carried out using collaboration and consensus with many public and private interests in Montana.

PUBLICATIONS: A list of publications, including titles on recycling and solid wastes, is available from the bureau.

Montana Environmental Information Center (MEIC)
P.O. Box 1184
Helena, MT 59624-1184
(406) 443-2520
(406) 443-2507 (FAX)
In-state recycling hotline: 1-800-823-MEIC
Anne Hedges, Program Director

MEIC is a member-supported, statewide organization dedicated to protecting and restoring Montana's environment. As such, it runs a recycling hotline to provide information on community recycling activities, locations of recycling centers, retailers that sell recycled products, and any other questions specifically related to recycling. The hotline is designed to provide free information and increase public awareness about source reduction and recycling. Founded in 1973, MEIC serves as an advocacy and research organization for the state's environmental issues, educating the public and providing citizens with organizational and technical assistance.

PUBLICATIONS: MEIC publishes and distributes several fact sheets on a variety of subjects, such as backyard composting, setting up an office recycling program, and cutting down on household hazardous waste. It also distributes two regular publications to its members: *Down to Earth,* a quarterly newsletter on Montana's environment, and *Capitol Monitor,* a biweekly legislative bulletin published during legislative sessions.

Nebraska

Nebraska Department of Environmental Quality (NDEQ)
Integrated Waste Management Section
Grants and Planning Unit
P.O. Box 98922
Lincoln, NE 68509-8922
(402) 471-4210
(402) 471-2909 (FAX)
Dannie Elwood, Environmental Program Specialist

"Since its inception in 1979, the Litter Reduction and Recycling Grant Program has provided nearly $9 million in assistance to programs and activities which reduce litter, provide litter reduction and recycling education, and promote recycling." Programs funded include projects to create products from recycled materials, establish a statewide recycling-marketing database, purchase recycling equipment, collect waste oil, and a variety of educational activities based on completing-the-cycle information.

PUBLICATIONS: The Grants and Planning Unit publishes an annual report to the governor and teams up with the Nebraska State Recycling Association to produce the *Nebraska Recycling Directory.* A guidance document, *Measuring and Tracking Recyclables and Organics,* and a booklet, *Keep Nebraska Beautiful,* are also available through this unit of the NDEQ, as is the brochure, *Market Development Opportunities in Recycling.*

Nebraska State Recycling Association (NSRA)
1941 S. 42nd Street, Suite 512
Omaha, NE 68105
(402) 444-4188
(402) 444-3953 (FAX)
In-state toll-free hotline: 1-800-248-7328
Kay Stevens, Executive Director

Founded in 1980, NSRA is a private, nonprofit membership organization "dedicated to serving the citizens and businesses of Nebraska in recycling and waste reduction issues." Its mission is to "serve as a catalyst: improving the consumer's ability to conserve resources by meeting solid waste management and recycling challenges with local resources, bridging the public/private gap, and helping to create local and regional markets for recycled products profitably." The association offers a toll-free hotline, runs a state recycling conference each fall, provides educational materials for youth groups and schools, assists in technical research, promotes recycling market development, and presents and distributes a variety of informative articles and papers.

PUBLICATIONS: NSRA publishes a directory of the state's recyclers and a monthly newsletter, the *Recycling Sentinel.* It also contributes to and produces the newsletter *Cooperative Connections,* a quarterly publication produced by seven recycling organizations from around the United States through a grant from the U.S.

Environmental Protection Agency. A package of fact sheets on a variety of waste issues, from recycling and composting to sustainable development and marketing services, is also available from NSRA.

Nevada

Nevada Division of Environmental Protection (NDEP)
333 W. Nye Lane
Carson City, NV 89706
(702) 687-4670 x3008
(702) 687-5856 (FAX)
Suzanne Sturtevant, Recycling Coordinator

This office serves as a state point of contact on recycling and coordinates efforts between counties. Recycling programs, if they exist, are administered at the county level. NDEP encourages recycling and administers grant programs serving that purpose.

New Hampshire

New Hampshire Department of Environmental Services (NHDES)
Waste Management Division
6 Hazen Drive
P.O. Box 95
Concord, NH 03302-0095
(603) 271-2901
(603) 271-2456 (FAX)
WWW: http://www.state.nh.us/des/descover.html
Sherry Godlewski, Recycling Coordinator

NHDES provides nonregulatory technical assistance to communities, districts, schools, and businesses throughout the state. Such assistance includes planning integrated solid waste management programs, recycling market development, and economic analysis of solid waste management programs. The department also tries to stimulate new recycling options and ideas, as well as the expansion of existing recycling programs. It encourages waste reduction, composting, and buying recycled through public education programs. NHDES also has a regulatory arm that implements the solid waste regulations and compliance standards for New Hampshire.

Northeast Resource Recovery Association (NRRA)
P.O. Box 721
Concord, NH 03302-0721
(603) 224-6996
(603) 226-4466 (FAX)
Spencer Bennett, Executive Director

Formerly known as the New Hampshire Resource Recovery Association, NRRA is a tax-exempt, nonprofit organization that provides marketing, educational, and technical services in waste reduction, recycling, composting, and other issues dealing with solid waste management. Its mission, in part, is "to provide leadership and support in promoting environmentally sound and economically viable waste reduction and recycling solutions." The association manages a cooperative marketing program, organizes regional workshops, and presents an annual conference and exposition.

PUBLICATIONS: NRRA publications include: *Recycling in NH: An Implementation Guide, Waste Plan—A Solid Waste Planning Model,* and a pamphlet entitled *Recycling in New Hampshire: A Beginner's Guide,* which includes a directory to the state's recycling centers. NRRA also produces a quarterly newsletter, *ReSource.*

New Jersey

Association of New Jersey Recyclers (ANJR)
120 Finderne Avenue
Bridgewater, NJ 08807
(908) 722-7575
(908) 722-8344 (FAX)
E-mail: anjr@ifu.net
Marie Kruzan, Director

ANJR is a nonprofit network, composed of members from both the private and public sectors, whose mission is "to serve as the voice of recycling in New Jersey through education and advocacy and the promotion of professional standards." This association promotes an integrated solid waste management plan through recycling, source reduction, composting, and reduction of household hazardous waste. The association coordinates roundtable discussions, equipment shows, and symposiums on solid waste issues, and designed one of the first online resource boards providing information on recycling and marketing.

PUBLICATIONS: Several publications are available from ANJR, including *Small Business Guide to Cost-Effective Recycling.*

New Jersey Department of Environmental Protection (NJDEP)
Bureau of Recycling and Planning
401 E. State Street, 2nd Floor
P.O. Box 414
Trenton, NJ 08625-0414
(609) 984-3438
(609) 777-0769 (FAX)
Guy Watson, Bureau Chief

NJDEP's goals include implementing the state's solid waste policy guidelines, which were written in response to former Governor James Florio's Emergency Solid Waste Task Force's final report/recommendations. As of 1987, NJDEP was directed by executive order to "develop a program maximizing source reduction and recycling and fashion a sound statewide planning process for managing our waste residues." Since then, NJDEP has worked cooperatively with other state agencies and districts to achieve a variety of goals, among them: the design of strategies to attain, "at least a 60 percent overall recycling rate by 1995 and 50 percent MSW [municipal solid waste] recycling rate," implementing aggressive source reduction programs, and creative planning and intergovernmental dialogue to achieve greater regionalization of recycling and waste disposal facilities. New Jersey did achieve its 1995 goal of 60 percent of the total solid waste stream, while reaching a recycling rate of 42 percent of municipal solid waste. Through its state recycling tax of $1.50 per ton of solid waste disposed at a landfill or transfer station, the state has generated sufficient funds to provide grants for five different categories relating to solid waste. Ten percent of this fund is specifically directed toward public information and eduction programs.

PUBLICATIONS: A list of publications dealing with recycling, composting, and source reduction is available from NJDEP.

New Mexico

New Mexico Environment Department
Solid Waste Bureau
Harold Runnels Building
1190 St. Francis Drive
Santa Fe, NM 87502
(P.O. Box 26110)
Santa Fe, NM 87502-6110
(505) 827-2883
(505) 827-2902 (FAX)
Cathy Tyson, State Recycling Coordinator

The mission of the Environment Department is to "preserve, protect, and perpetuate New Mexico's environment for present and future generations. In the case of solid waste, this is accomplished through programs which assist local governments, civic organizations, businesses, and individual citizens in their efforts to reduce the amount of waste generated, [and] to reuse and recycle as much potential waste as possible."

PUBLICATIONS: The Outreach Section of the bureau publishes a quarterly newsletter entitled *Focus on Solid Waste,* which is distributed "as a service for those who have an interest in solid waste. We will keep you in touch with issues, innovative technologies, and new regulations that relate to solid waste practices." A brochure, "Backyard Composting," a display for use at fairs and conventions, several elementary education materials, as well as videotapes, slides, teaching aids, and fact sheets are also available through the Solid Waste Bureau.

New Mexico Recyclers Association
University of New Mexico
1211 University Boulevard, NE
Albuquerque, NM 87131-3061
(505) 277-1681
(505) 277-3226 (FAX)
Jim Davis, Director

The New Mexico Recyclers Association is a nonprofit organization devoted to promoting recycling throughout the state.

PUBLICATIONS: The organization publishes a *State Directory of Recyclers* each year, which is distributed to recyclers and government agencies "to help keep lines of communication, regarding recycling, open between citizens, business, and lawmakers in New Mexico."

New York

New York State Association for Reduction, Reuse and Recycling (NYSAR³)
51 Fulton Street
Poughkeepsie, NY 12601
(518) 463-7964
(518) 463-2560 (FAX)
WWW: http://www.recycle.net/recycle/assn/nysarrr
Eileen McGuire, President

NYSAR³ is a nonprofit organization of diverse membership whose members are united by a common goal to "utilize waste reduction, reuse, and recycling techniques to the fullest extent possible in order to achieve the benefits associated with these efforts, including, but not limited to, comprehensive waste management strategies, natural resource conservation, environmental protection, energy conservation, as well as social and economic development." NYSAR³ provides forums for exchange of information, a clearinghouse of information on reduction, reuse, and recycling, legislative information with regard to waste issues, and feedback to government agencies. NYSAR³ is a member of the National Recycling Coalition.

PUBLICATIONS: NYSAR³ produces a quarterly newsletter and maintains a clearinghouse of waste-related information.

New York State Department of Environmental Conservation (NYSDEC)
Waste Reduction and Recycling
50 Wolf Road
Albany, NY 12233-4015
(518) 457-7337
(518) 457-1283 (FAX)
Norman H. Nosenchuck, Director, Division of Solid Waste

NYSDEC is the state regulatory agency that deals with recycling and other solid waste issues. As such, it was responsible for writing the State Solid Waste Management Plan, which was first issued in 1987. One goal outlined in this plan called for reducing the waste stream by 50 percent by 1997 through the 3Rs of reduce, reuse, and recycle (as of April 1996, New York had attained a recycling rate of 32 percent).

PUBLICATIONS: Copies of the plan and more than 20 other publications are available at no cost through NYSDEC, as is a list of educational curricula and videos.

North Carolina

North Carolina Division of Waste Management (NCDWM)
401 Oberlin Road, Suite 150-254
Raleigh, NC 27605
(919) 733-0692
(919) 733-4810 (FAX)
WWW: http://www.watenot.ehnr.state.nc.us/
Paul Crissman, Recycling Contact

The NCDWM's mission is to "ensure proper management of solid and hazardous waste through waste reduction, sound recycling methods, safe management practices, and proper waste disposal."

PUBLICATIONS: A list of publications on solid waste and recycling is available either online (see WWW address) or by calling (919) 733-0692.

North Carolina Recycling Association (NCRA)
7330 Chapel Hill Road, Suite 207
Raleigh, NC 27607-5042
(919) 851-8444
(919) 851-6009 (FAX)
WWW: http://www.recycle.net/recycle/ncra/index.html
E-mail: ncrecycles@aol.com
Craig Barry, Executive Director

Founded in 1988, NCRA promotes resource conservation by advancing waste reduction and recycling efforts across the state. NCRA is a membership organization composed of government and private industry representatives and individuals involved with recycling. Their goal is to facilitate the development of the recycling industry and to assist recyclers by offering practical advice and solutions to help them achieve their solid waste management goals.

PUBLICATIONS: NCRA publishes a newsletter, *The R-Word,* on a quarterly basis, as well as a resource guide for businesses, a *Green Building Products Directory,* and a membership directory.

North Dakota

North Dakota Department of Health and Consolidated Laboratories (NDDHCL)
Solid Waste Management Program
1200 Missouri Avenue
P.O. Box 5520
Bismarck, ND 58502-5520
(701) 328-5166
(701) 328-5200 (FAX)
WWW: http://165.234.109.13/ndhd/index.html
Robert Tubbs-Avalon, State Recycling Coordinator

The Solid Waste Management Program of NDDHCL regulates the storage, collection, transportation, and disposal of municipal solid wastes to preserve and enhance water and land resources and to protect public health and property. In 1991, legislation was passed declaring goals for the reduction of landfilled solid waste. This bill also directed the department to provide public education in the areas of waste reduction, source separation, reuse, recycling, and appropriate management of solid waste. Such programs have been developed and implemented, resulting in a 28 percent statewide recycling rate in 1996 (the state's goal is 40 percent by the year 2000). The state's efforts have also created 400 full- and part-time recycling jobs within the state and allowed at least half of the state's population access to recycling: 90,000 are served by curbside programs, while another 200,000 have municipal drop-off programs available for their use.

PUBLICATIONS: Pamphlets on how to dispose of or safely deal with Household Hazardous Waste are available through the NDDHCL, and a brochure entitled *Recycling Works!* was due out in August 1997. An annual Recycling Directory is available through NDDHCL, as are resources dealing with the broader issues of environmental education.

Ohio

Association of Ohio Recyclers (AOR)
P.O. Box 242
Kirkersville, OH 43033
(614) 927-3200
(614) 927-1147 (FAX)
E-mail: pjsmith@prgone.com
Mary Wiard, Director

Founded in 1989, this nonprofit trade organization's mission is "to promote recycling and waste prevention while serving our members through information exchange, education, networking, and advocacy." AOR sponsors seminars, workshops, and facility tours, and is a member of the National Recycling Coalition.

PUBLICATIONS: Members receive both the AOR and the NRC newsletters on a bimonthly basis.

Ohio Department of Natural Resources
Division of Recycling and Litter Prevention (DRLP)
1889 Fountain Square, Building F-2
Columbus, OH 43224
(614) 265-7069
(614) 262-9387 (FAX)
Toll-free hotline: 1-800-317-4797
WWW: http://www.dnr.ohio.gov/odnr/recycling
Jenni Worster, Division Chief

DRLP's mission is to "provide leadership and financial support to advance waste reduction, recycling, and litter prevention programs that positively impact the citizens of Ohio." As such, the DRLP is responsible for administering grants for recycling projects, promoting recycling through educational and public awareness programs, and providing advisory and technical assistance to recycling, waste reduction, recycling market development, and litter prevention programs around the state. The division also assists in implementing Ohio's Solid Waste Management Plan. Through the Recycle, Ohio! program, the division develops model recycling programs designed to establish new and expand existing recycling initiatives throughout the state. Educational programs run by DRLP include Super Saver Investigators (K-6), Investigating Solid Waste Issues

(7-12), Ohio Recycle Month, a recycling exhibit at the Ohio Center of Science and Industry, a Governor's Award for student research, a science workbook on litter prevention and recycling, a Girl Scout merit patch, and "Recycle with Ohio Zoos." DRLP also runs an office paper recycling program called PAPERCYCLE and has worked to encourage the state's procurement of products made from recycled materials.

PUBLICATIONS: Published materials include *Super Saver Investigators,* a 400-page elementary school guide about solid waste and recycling; *Ohio Science Workbook: Litter Prevention and Recycling; Recycling Basics; Directory of Ohio Recyclers; PAPERCYCLE: A Guide to Office Paper Recycling;* and *Recycled Products Guide,* as well as various studies on recycling and solid waste. Numerous fact sheets are available from DRLP, as well as a quarterly newsletter, *The Recycle, Ohio! Communique.*

Oklahoma

Oklahoma Department of Environmental Quality (ODEQ)
Waste Management Division
1000 NE 10th Street
Oklahoma City, OK 73117-1212
(405) 271-5338/(405) 271-1400
(405) 271-8425 (FAX)
WWW: http://www.deq.state.ok.us/waste.html
Bruce Hulsey, General Recycling Coordinator

Oklahoma Recycling Association (OKRA)
P.O. Box 26911
Oklahoma City, OK 73126-6911
(405) 945-1932
Michelle Pitt, Executive Director

OKRA was incorporated in 1996 as a nonprofit corporation and is still in the process of getting organized. Its mission statement is: "To improve the business of recycling in Oklahoma." OKRA works with individual counties to develop solid waste management plans and is working to educate the public on the entire cycle of recycling, from collection to the purchasing of recycled materials.

PUBLICATIONS: OKRA publishes a bimonthly newsletter, an annual environmental guide, and a recycling directory/guide.

Oregon

Association of Oregon Recyclers (AOR)
P.O. Box 483
Gresham, OR 97030-0107
(503) 661-4475
(503) 524-2373 (FAX)
E-mail: aor@mindspring.com
Charlotte Becker, Executive Director

AOR is "a nonprofit organization of recycling professionals and activists committed to reducing waste and improving recycling in Oregon." AOR has "furthered recycling in Oregon through networking assistance and information exchange among recyclers; advocating improved market conditions and prices; promotions and education about recycling; and promoting legislation for recycling." The association also holds a yearly conference.

PUBLICATIONS: AOR publishes a membership directory and a monthly newsletter.

Oregon Department of Environmental Quality (ODEQ)
811 SW Sixth Avenue
Portland, OR 97204-1390
(503) 229-6823
(503) 229-6954 (FAX)
In-state toll-free hotline: 1-800-452-4011
Joan Grimm, Waste Reduction Education Specialist

ODEQ oversees the state's recycling programs, dealing with the public, communities, and other large generators and manufacturers of solid waste. It is primarily concerned with legislation, education, and economics and with promoting the effectiveness of its programs. The department has also developed a recycling curriculum (known as *Rethinking Recycling*) that has been widely distributed and utilized in state schools.

PUBLICATIONS: Several publications, curricula guides, and audiovisual materials are available from ODEQ in conjunction with the *Rethinking Recycling* education program.

Pennsylvania

Commonwealth of Pennsylvania Department of Environmental Resources (DER)
Bureau of Waste Management
Division of Waste Minimization and Planning
P.O. Box 8471
Harrisburg, PA 17105-8471
(717) 787-7382
(717) 787-1904
Recycling hotline: 1-800-346-4242
WWW: http://www.dep.state.pa.us/dep/
Keith Kerns, Division Chief

DER oversees compliance with the state's Municipal Waste Planning, Recycling, and Waste Reduction Act (Act 101), and it administers grants in conjunction with this legislation. Additionally, the department has developed a school curriculum on waste reduction and recycling, and it provides speakers, sponsors conferences, and offers technical assistance.

PUBLICATIONS: DER publishes a series of fact sheets on solid waste, including one on recycling information and technical services. The department also provides a list of free publications.

Pennsylvania Resources Council (PRC)
3606 Providence Road
Newtown Square, PA 19073
(610) 353-1555
(610) 353-6257
In-state toll-free hotline: 1-800-GO-TO-PRC
WWW: http://www.voicenet.com/~rusw/prc/
E-mail: imperato@prc.org
Pat Imperato, Executive Director

PRC was founded in 1939 and was originally called the Pennsylvania Roadside Council. It is the oldest citizen-action environmental organization in the state and currently works with a wide range of groups that implement recycling and waste reduction programs. PRC provides recycling and waste reduction expertise, community education publications and programs, conferences, seminars, videos, and materials for educators. Additionally, it runs several environmental programs, including the Environmental Shopping program, a Buy Recycled Campaign, and the Business for a Green America program. Technical information and assistance are also available through PRC on such subjects as waste reduction, recycling, litter control, and beautification. *PenCycle* is a statewide computer information network and bulletin board founded in 1989 and run by PRC. Most of the information and services in *PenCycle* are in the process of being transferred to PRC's World Wide Web page.

PUBLICATIONS: An extensive list of publications, games, and promotional materials on subjects ranging from litter control to recycling and composting is available from PRC. Three newsletters are published on a quarterly basis: *PRC Newsletter, All about Recycling,* and *Environmental Shopping Update.*

Rhode Island

Rhode Island Department of Environmental Management
Office of Strategic Planning and Policy
OSCAR Program
235 Promenade Street
Providence, RI 02908
(401) 277-3434
(401) 277-2591 (FAX)
In-state toll-free recycling hotline: 1-800-CLEANRI
E-mail: oscar@oscar.state.ri.us
Terri Bisson, Program Manager—Municipal Recycling
Marty Davey, Program Manager—Commercial Recycling

OSCAR works in partnership with the Solid Waste Management Corporation, the Rhode Island Department of Administration, municipalities, and businesses to reduce the amount of waste placed in state landfills. Rhode Island's mandatory recycling program is coordinated by the state and applies to both homeowners and big business.

PUBLICATIONS: This organization's literature includes *Recycling in Rhode Island: A Blueprint for Success, Hazardous Wastes from Homes,* and *We Recycle: Getting Started with Recycling. OSCAR's CALL* is a newsletter that is "published on a regular basis."

South Carolina

South Carolina Department of Health and Environmental Control (SCDHEC)
Office of Solid Waste Reduction and Recycling (OSWRR)
2600 Bull Street
Columbia, SC 29201
(803) 896-4232
(803) 896-4001 (FAX)
Toll-free recycling hotline (nationwide): 1-800-768-7348
Bill Culler, Director of Solid Waste Management

The SCDHEC was created by the South Carolina Solid Waste Policy and Management Act of 1991 and was directed by the governor and general assembly to develop a state solid waste management plan by 27 November 1992 that would "reflect serious deliberation not given to energy and environmental issues before in this State." This first plan, when completed, evaluated the then-current solid waste practices, inventoried and categorized the wastes generated, and estimated future trends and the state's capacity to manage its solid wastes. Subsequently, a Solid Waste Advisory Council and a Recycling Market Development Council were set up. The SCDHEC works with these councils and oversees the regional and county solid waste management programs that were created by plans submitted in 1994. It works toward the goals established by the Act to "reduce the amount of solid waste received at municipal solid waste landfills and incinerators by 30 percent, calculated by weight" and to "recycle at least 25 percent, calculated by weight, of the total solid waste stream generated in this state" by 27 May 1997. As of 1995, the state's recycling rate had reached 16 percent.

PUBLICATIONS: SCDHEC will provide "An Overview of the S.C. Solid Waste Policy and Management Act of 1991" on request.

South Carolina Recycling Association (SCRA)
P.O. Box 11937
Columbia, SC 29211
(803) 771-4271
(803) 771-4272 (FAX)
Joe S. Jones, Executive Director

Founded in 1989, SCRA is a nonprofit organization established to promote recycling through education. The association sponsors workshops and seminars throughout the year, and maintains an informative resource library. SCRA is a member of the National Recycling Coalition.

PUBLICATIONS: SCRA produces a newsletter, *Full Circle,* at least twice a year, and its resource library contains video and audiotapes, reference lists, curriculum guides, and an assortment of brochures.

South Dakota

South Dakota Department of Environment and Natural Resources
Waste Management Program
523 E. Capitol Avenue
Pierre, SD 57501-3182
(605) 773-6498
In-state toll-free hotline: 1-800-848-8203
WWW: http://www.state.sd.us/state/executive/denr/denr.html
Terry Keller, Director, Waste Management Program

The Office of Waste Management is responsible for the department's regulatory efforts in the areas of solid waste, hazardous waste, asbestos, lead abatement, PCBs, and recycling. In addition, the office coordinates the state's solid waste recycling efforts.

South Dakota Solid Waste Management Association (SDSWMA)
Recycling Branch
P.O. Box 957
Rapid City, SD 57709-0957
(605) 394-6747
(605) 394-2497 (FAX)
Deborah A. Barton, Executive Director

SDSWMA was incorporated as a nonprofit organization in September 1991. It is comprised of members from recycling, collection, and waste disposal interests. The organization promotes "environmentally sound solid waste management practices" and provides an information network for its membership, as well as its six surrounding states, where applicable. Solid waste management training and seminars, technical information and assistance, and research support are other services offered to members.

PUBLICATIONS: The SDSWMA newsletter, *Material Matters,* is published on a quarterly basis.

Tennessee

Tennessee Department of Environment and Conservation (TDEC)
Division of Solid Waste Assistance (DSWA)
401 Church Street, 14th Floor
Nashville, TN 37243-0455
(615) 532-0091
(615) 532-0231 (FAX)
Paul Evan Davis, Director of Solid Waste Assistance

DSWA was established by the state's Solid Waste Management Act of 1991 in order to "implement the activities mandated by the law and to help local governments receive grants and services," which are provided from a dedicated Solid Waste Management Fund. This fund is supported by a surcharge on each ton of waste disposed and by a predisposal tire fee. DSWA has three sections: Grants Administration, Special Wastes, and Recycling. The Recycling Section provides special assistance to local governments in developing recycling programs and markets.

PUBLICATIONS: Two pamphlets are published by TDEC for state offices: *The Whys and Hows of Recycling at the Office* and *It's Getting Easier!*

Tennessee Recycling Coalition (TRC)
1250 E. 3rd Street
Chattanooga, TN 37404
(615) 697-1408
(615) 757-5350
Bill Penn, President

TRC is a nonprofit association consisting of a membership consensus from public, private, and nonprofit environmental organizations. It presents a conference on recycling within the state each year.

PUBLICATIONS: TRC produces the *Tennessee Recycling Report* on a quarterly basis. An educational brochure on recycling is being developed by the membership at this time.

Texas

Recycling Coalition of Texas (RCT)
P.O. Box 2359
Austin, TX 78768-2359
(512) 469-6079
(512) 472-7026
Ben Walker, President

Founded in 1989, RCT is a nonprofit alliance of recyclers from around the state that cosponsors recycling conferences and seminars with Texas A&M University and encourages recycling and sound solid waste practices throughout the state. The RCT promotes waste reduction through public policy advocacy, professional networking, and public education "in order to conserve finite natural resources and protect and enhance our natural environment within a sustainable framework." RCT is a member of the National Recycling Coalition, and maintains both Technical and Policy Councils to advise and assist both industry and government.

PUBLICATIONS: A newsletter, *Inside Texas,* is published on a quarterly basis by RCT.

Texas Natural Resource Conservation Commission (TNRCC)
Office of Pollution Prevention and Recycling (OPPR)
P.O. Box 13087
Austin, TX 78711-3087
(512) 239-6750
(512) 239-6763 (FAX)
In-state toll-free information center: 1-800-64-TEXAS
WWW: http://www.tnrcc.state.tx.us
Andrew Neblett, Division Director of OPPR

TNRCC strives to protect the state's "precious human and natural resources consistent with sustainable economic development." Its goal is clean air and water and safe management of waste, with an emphasis on pollution prevention. The commission is committed "to providing efficient, prompt, and courteous service to the people of Texas, ever mindful that our decisions must be based on common sense, good science, and fiscal responsibility."

PUBLICATIONS: TNRCC produces a publications catalog that includes over 20 titles devoted specifically to recycling issues. A copy of this catalog may be obtained by calling (512) 239-0028 or on the World Wide Web at http://www.tnrcc.state.tx.us/admin/topdoc/pubat.html.

Utah

Salt Lake Valley Solid Waste Management Facility (SLVSWMF)
Recycling Information Office
6030 W. 1300 South
Salt Lake City, UT 84104
(801) 974-6902
(801) 974-6905
Margaret Grochocki, Public & Recycling Information Officer

SLVSWMF is associated with the Salt Lake Valley Landfill. It promotes alternatives to the landfill disposal of solid waste.

PUBLICATIONS: SLVSWMF provides a list of pamphlets on recycling, composting, grass recycling, household hazardous waste, and source reduction. Other materials, such as videotapes, books, and periodicals, are available for loan. The facility also distributes an annual recycling guide/directory.

Utah Department of Environmental Quality (UDEQ)
168 N. 1950 West
Salt Lake City, UT 84114-4810
(801) 536-4477
(801) 536-0061 (FAX)
WWW: http://www.eq.ex.state.ut.us/eqoas/p2/p2_home.htm
Sonja F. Wallace, Recycling Coordinator

Utah's Recycling Task Force was established by House Bill 334 (Waste Recycling Task Force) in 1991. This task force issued a report to the legislature in December of that year, recommending more state action with regard to recycling, source reduction, and use of recycled materials. The UDEQ's current focus is on the development of local markets, and to that end Utah recently instituted statewide market development zones, which provide tax benefits for recycling companies.

Vermont

Association of Vermont Recyclers (AVR)
P.O. Box 1244
Montpelier, VT 05601-1244
(802) 229-1833
(802) 229-9408 (FAX—call first)
WWW: http://www.sover.net/~recycle/index.html
E-mail: recycle@sover.net
Barry Lampke, Executive Director

Founded in 1982, AVR is a nonprofit organization whose members "promote solid waste reduction and recycling." Its goals are to "promote recycling and waste reduction in Vermont through education and cooperative efforts, to provide an objective resource and referral base for individual municipalities, schools, businesses and solid waste districts, and to encourage the increased use of products made with recycled materials." AVR sponsors an annual conference, offers recycling seminars and workshops, and provides technical assistance. It is a member of the National Recycling Coalition.

PUBLICATIONS: AVR publishes the quarterly newsletter *Beyond the Bin* and can provide resource packets for various business types. AVR also publishes a *Business Recycling Manual* and *Closing the Loop—A Guide to Buying Recycled Products.* Additionally, the association produces a variety of materials on recycling, waste reduction, and composting, as well as numerous teaching aids, posters, and other resources, including the VHS video, *Taking It to the Depot.*

Vermont Department of Environmental Conservation (VDEC)
Environmental Assistance Division (EAD)
103 S. Main Street
Laundry Building
Waterbury, VT 05671-0411
(802) 241-3589
(802) 241-3273 (FAX)
Marci Towns, Recycling Specialist

EAD is the part of VDEC that deals specifically with recycling. Its 1992 work plan identified six goals: (1) to reduce the amount of unregulated hazardous waste being disposed of; (2) to educate the public about recycling, reduction, and hazardous waste; (3) to increase local markets for recyclables; (4) to provide technical assistance to Vermont municipalities, solid waste districts, and businesses; (5) to increase composting of high-quality organic wastes; and (6) to reduce the amount of state government waste requiring disposal and increase the purchase of materials with recycled content.

PUBLICATIONS: This section has published numerous reports, brochures, and guidance documents on recycling, environmental shopping habits, and composting that are available to state residents for free; a small fee is charged for out-of-state orders.

Virginia

Virginia Department of Environmental Quality (VDEQ)
P.O. Box 10009
Richmond, VA 23240
(804) 698-4447
(804) 698-4453 (FAX)
WWW: http://www.deq.state.va.us
Larry Lawson, Environmental Improvement Director

The VDEQ's mission is to "protect the environment of Virginia in order to promote the health and well-being" of its citizens. To accomplish these goals, the VDEQ administers state and federal environmental programs, issues environmental permits and ensures compliance with regulations, and coordinates planning between the state's variety of environmental programs. Virginia's DEQ is in the process of developing and implementing a waste tire program, which will include strategies for dealing with the tire stockpiles around the state. The department also administers an annual grant program that provides funds for local litter prevention and recycling efforts.

PUBLICATIONS: The VDEQ publishes *25 Ways to Help Virginia's Environment*, and an extensive list of publications is available through the Library of Virginia by calling (804) 692-3592 or on the World Wide Web at http://www.vsla.edu/.

Virginia Recycling Association (VRA)
2210 Mt. Vernon Avenue
Alexandria, VA 22301
(703) 549-9263
(703) 549-9264 (FAX)
Ron Nitzche, Executive Director

Washington

Washington State Department of Ecology (WDE)
Solid Waste Services and Financial Assistance Program
P.O. Box 47600
Olympia, WA 98504-7600
(360) 407-6000
(360) 407-6102
In-state toll-free hotline: 1-800-RECYCLE (1-800-732-9253)
WWW: http://www.wa.gov/ecology
Jim Pendowski, Program Director

WDE was created in 1970 to promote educational programs and enforce the State Environmental Policy Act. Since then, it has created recycling and waste reduction programs for offices and businesses, conducted studies on solid waste management alternatives, and established public education programs throughout the state. The department sponsors an Ecology Youth Corps for 14- to 17-year-olds to teach the importance of waste reduction, recycling, and litter control.

PUBLICATIONS: WDE publishes many brochures and fact sheets on recycling and litter control, composting, and recycled paper products. It also publishes materials on household hazardous wastes and a waste management curriculum for kindergarten through twelfth grade. Educational videos, brochures, and posters can be ordered through the statewide hotline. More information may be found on WDE's World Wide Web page.

Washington State Recycling Association (WSRA)
6100 Southcenter Boulevard, Suite 180
Tukwila, WA 98188-2486
(206) 244-0311
(206) 244-4413 (FAX)
E-mail: wsra@aol.com
Dot Vali, Executive Director

WSRA is the professional trade organization for the recycling industry in Washington State. The association "has been active in providing recycling information to the general public, establishing and maintaining standards for the recycling industry, and providing a forum for its members to exchange information." It also holds an annual conference/trade show.

PUBLICATIONS: WSRA publishes a monthly newsletter.

West Virginia

West Virginia Division of Environmental Protection (DEP)
Solid Waste Management Section
1356 Hansford Street
Charleston, WV 25301
(304) 558-6350
(304) 558-1574 (FAX)
Paul Benedum, Environmental Resource Specialist

West Virginia's DEP has developed and distributed a curriculum on solid waste management for kindergarten through twelfth-grade classes. The division assists individuals, communities, businesses, and industry in the common objective of meeting the legislated goal of a 30 percent reduction in the state's waste stream by the year 2000 through source reduction, recycling, reuse, and resource recovery.

PUBLICATIONS: DEP publishes several pamphlets on recycling.

Wisconsin

Associated Recyclers of Wisconsin (AROW)
P.O. Box 44008
Madison, WI 53744-4008
(608) 277-1978
(608) 277-9266 (FAX)
E-mail: cow@mailbag.com
Bob Jordan, President

AROW is a nonprofit, educational coalition "dedicated to protecting the environment and preserving natural resources through waste reduction, reuse, recycling, and composting." AROW provides networking through the state recycling programs, works on developing and updating state recycling policies, and is a member of the

National Recycling Coalition. The association regularly presents various regional workshops, as well as an annual conference.

PUBLICATIONS: AROW publishes a newsletter five times a year.

Wisconsin Department of Natural Resources
Bureau of Cooperative Environmental Assistance
Box 7921
101 S. Webster
Madison, WI 53707-7921
(608) 267-0873
(608) 267-0496 (FAX)
Shelley Heilman, Environmental Outreach Coordinator

PUBLICATIONS: The department has a list of approximately 100 free publications on recycling/solid waste, including educational materials for kindergarten through twelfth grade, that are available to the public.

Wyoming

Wyoming Department of Environmental Quality (WDEQ)
Solid & Hazardous Waste Division
250 Lincoln Street
Lander, WY 82520
(307) 332-6924
(307) 332-7726 (FAX)
WWW: http://deq.state.wy.us/ms.htm
E-mail: dhogle@missc.state.wy.us
Diana Gentry-Hogle, State Recycling Coordinator

In January 1992, the Wyoming Governor's Committee on Recycling presented its recommendations to the governor. Shortly thereafter, the state created the position of recycling coordinator, responsible for providing communities with "direct, functional technical assistance regarding market and transportation information." The coordinator is also directed to gather information on cooperative marketing between communities and nonconventional market sources with regard to recycling, and to work "with the public to encourage and enhance waste reduction and recycling opportunities throughout Wyoming."

PUBLICATIONS: A copy of the committee's report, a Wyoming Recycling Directory, *A Guide to Backyard Composting in Wyoming,* and a pamphlet entitled *Do It!* are available through WDEQ. Other literature is being planned, as are educational programs for the public.

Wyoming Recycling Association (WRA)
P.O. Box 539
Laramie, WY 82070
(307) 632-1245
(307) 777-3773
Kristin Span, Executive Director

Founded in 1992, WRA's mission is "to promote recycling and assist with cooperative marketing of recycled resources that would otherwise be landfilled or incinerated." WRA provides technical assistance, educational presentations, a speakers bureau, workshops, and an information clearinghouse to members and state residents.

PUBLICATIONS: A newsletter, the *Wyoming Re*Cycler,* is published bimonthly. A recycling and waste reduction library is maintained by WRA, and an *Integrated Solid Waste Management Handbook* is available for Wyoming communities.

National Organizations and Federal Government Agencies

Aluminum Association, Inc. (AA)
900 19th Street, NW, Suite 300
Washington, DC 20006
(202) 862-5163
(202) 862-5164 (FAX)
David N. Parker, President

AA is the trade association for domestic producers of primary (virgin) and secondary (recycled) aluminum and semifabricated products (items prepared by one manufacturer for another manufacturer to make into finished products). The association provides leadership to the industry through its programs and services, which aim to enhance aluminum's position in a world of proliferating materials, increase its use as the "material of choice," remove impediments to its fullest use, and assist in achieving the industry's environmental, societal, and economic objectives. Member companies operate 300 plants in 40 states.

PUBLICATIONS: AA publishes a list of its members, booklets entitled "Aluminum Pays—Aluminum Gets Recycled" and "Aluminum—The 21st Century Metal," both of which contain information on aluminum recycling and solid waste in the United States, as well as additional information on the various forms of aluminum used by individuals and industries. AA also publishes "Aluminum Update" fact sheets, which provide specific information about such topics as "Lightweighting" and "Automotive Aluminum."

American Forest & Paper Association, Inc. (AF&PA)
1111 19th Street, NW, Suite 800
Washington, DC 20036
1-800-878-8878
W. Henson Moore, President

In 1993, the American Paper Institute, the National Forest Products Association, and the American Forests Council combined forces to become the AF&PA, the national trade association of the forest, pulp, paper, paperboard, and wood products industry. Its members account for more than 90 percent of domestic paper and recycled paper manufacturing capacity. The association's leadership is spearheading an industry-wide effort to establish environmental principles for forest management, mill operations, and sound recovery/recycling/reuse programs for wood, paper, and other forest products.

PUBLICATIONS: The AF&PA publishes "Papermatcher," a directory of paper recycling mills; "How to Recover Paper for Recycling," a guide for corporations, retailers, municipalities, and individuals; "Recovered Paper Statistical Highlights"; and a poster on paper recycling.

American Plastics Council (APC)
1801 K Street, NW
Washington, DC 20006
(202) 974-5400
(202) 296-7119 FAX
Plastics hotline: 1-800-243-5790
WWW: http://www.plasticsresource.com
Red Cavaney, President

APC is a national organization whose mission is to actively demonstrate that plastics are a preferred material and a responsible choice in a more environmentally conscious world. The council is a joint initiative with the Society of the Plastics Industry, Inc., and focuses on creating programs addressing resource conservation issues and product benefits. (The Council for Solid Waste Solutions was absorbed into this organization in 1992.)

PUBLICATIONS: APC publishes a number of brochures for the general public as well as other publications geared toward communities, schools, and corporations. A list of available materials can be obtained by calling the Plastics hotline.

Environmental Action Foundation (EAF)
1525 New Hampshire Avenue, NW
Washington, DC 20036
(202) 745-4879
WWW: http://www.econet.apc.org/eaf
E-mail: eaf@igc.apc.org
Ruth Caplan, Executive Director

EAF is part of Environmental Action, Inc., a national membership-based organization that works for strong state and federal environmental laws. Part of EAF is the Solid Waste Alternatives Project (SWAP), which promotes source reduction, recycling, and composting.

PUBLICATIONS: SWAP publishes a quarterly newsletter, *Wastelines,* and EAF publishes the bimonthly magazine, *Environmental Action*. EAF also has a list of 21 other publications available for purchase, many of which deal specifically with solid waste/recycling issues.

Environmental Defense Fund (EDF)
257 Park Avenue South
New York, NY 10010
(212) 505-2100
Public information hotline: 1-800-684-3322
WWW: http://www.edf.org
E-mail: webmaster@edf.org
Fred Krupp, Executive Director

EDF created the slogan, "If you're not recycling, you're throwing it all away." A large, nonprofit, national environmental organization, EDF is devoted to protecting the environment, mainly through legislative lobbying and lawsuits against major corporate offenders. It was EDF's work with McDonald's Corporation that led that company to discontinue its use of polystyrene and to establish its national incentive program to "buy recycled." EDF has six regional offices in addition to its New York location.

PUBLICATIONS: EDF publishes the *EDF Letter* on a monthly basis, and offers a catalog of publications on environmental issues. EDF has numerous pamphlets, studies, and reports in the subject areas of waste reduction, recycling, composting, and solid waste management in addition to life-cycle assessments of paper products.

Environmental Industry Associations (EIA)
4301 Connecticut Avenue, NW, Suite 300
Washington, DC 20008
(202) 244-4700
(202) 966-4818 (FAX)
WWW: http://www.envasns.org
Bruce Parker, President

EIA "represents about 2,000 individuals and companies that manage solid, hazardous, and medical wastes; manufacture and distribute waste equipment; provide environmental managing and consulting; and offer related pollution-prevention services." The organization is made up of four other associations: the National Solid Wastes Management Association, the Waste Equipment Technology Association, the Hazardous Waste Management Association, and the Society of Environmental Management and Technology.

PUBLICATIONS: EIA produces an *Environmental Education Resource Catalog,* which offers numerous books, pamphlets, and reprints on subjects ranging from landfill technology to recycling economics.

Flexible Packaging Association (FPA)
1090 Vermont Avenue, NW, Suite 500
Washington, DC 20005-4960
(202) 842-3880
(202) 842-3841 (FAX)
E-mail: fpa@ssn.com
Glenn E. Braswell, President

FPA is a trade association for manufacturers and suppliers of flexible packaging materials and allied products. Flexible packaging includes bags, pouches, labels, liners, and wraps made of plastic film, paper, aluminum foil, or a combination thereof. Such packaging is compressible or collapsible, having no real shape of its own without the product is holds. It generally has a thickness of 10 millimeters or less. The goals of the association include communicating with state and federal governments regarding the packaging industry, promoting the use of flexible packaging, and conducting research. The motto on their literature packet, "Source Reduction: Less Waste in the First Place," indicates the industry's emphasis on the first step in the U.S. Environmental Protection Agency's solid waste hierarchy.

PUBLICATIONS: Numerous publications, mostly technical, are available from FPA, as are several brochures and a poster that address source reduction and solid waste solutions. A newsletter, *FPA Update,* is published on a monthly basis.

Foodservice & Packaging Institute, Inc. (FPI)
1901 N. Moore Street, Suite 1111
Arlington, VA 22209
(703) 527-7505
(703) 527-7512 (FAX)
WWW: http://www.fpi.org/fpi
Joseph Bow, President

Founded in 1993, FPI is "the material-neutral trade association for manufacturers, raw material suppliers, machinery suppliers, and distributors of foodservice disposable products, as well as others associated with the industry." FPI promotes the "sanitary, safety, functional, economic, and environmental benefits of foodservice disposables," and supports the "environmentally responsible manufacture, distribution, use and disposal of these products."

PUBLICATIONS: Several informational pamphlets and fact sheets, a litter control checklist, and sanitation studies on disposables are available through the FPI office. FPI also distributes an information package, including its annual *Environmental Stewardship Report,* and videotape, *Foodservice Disposables: Should I Feel Guilty?*

Glass Packaging Institute (GPI)
1627 K Street, NW, Suite 800
Washington, DC 20006
(202) 887-4850
(202) 785-5377 (FAX)
WWW: http://www.gpi.org
Lewis Andrews, President

GPI is a trade association that represents over 90 percent of the glass container manufacturers in America. The association supports glass recycling efforts across the United States through consumer education, technical assistance to municipalities, and industry education and promotion.

PUBLICATIONS: GPI publishes several pamphlets and fact sheets (Issue Alerts) about glass recycling. Included in their pamphlet "Glass Recycling: Why? How?" is a directory of all domestic glass container manufacturing plants (a total of 79). *Glass Recycling,* the association's newsletter, is published quarterly.

INFORM, Inc.
120 Wall Street, 16th Floor
New York, NY 10005-4001
(212) 361-2400
(212) 361-2412 (FAX)
E-mail: Inform@igc.apc.org
Joanna D. Underwood, President

INFORM is a nonprofit environmental research and education organization that provides the public with information about the environmental impact of business and institutional practices. INFORM's goal is to "identify specific solutions—mostly preventive in nature—to increasingly complex environmental problems." The research, reports, and communications prepared by INFORM are focused on three main areas: sustainable products and practices, chemical hazards prevention, and energy and transportation. INFORM neither lobbies nor litigates.

PUBLICATIONS: The organization has published a number of books and reports on solid waste management, all of which are available directly through INFORM. Their newsletter, *INFORM Reports,* is published quarterly, and 14 free fact sheets are available to the public.

Institute for Local Self-Reliance (ILSR)
2425 18th Street, NW
Washington, DC 20009-2096
(202) 232-4108
(202) 332-0463 (FAX)
WWW: http://www.great-lakes.net:2200/0/partners/ILSR/ILSRhome.html
E-mail: ilsr@igc.apc.org
Neil Seldman, President

ILSR is a nonprofit educational and research organization. It works in urban areas to join "technical ingenuity with a sense of community to establish sustainable, environmentally sound forms of consumption and production." It promotes self-reliance in communities by "investigating examples of closed-loop manufacturing, materials policy, materials recovery, energy efficiency, and small scale production." ILSR helps communities throughout the United States and abroad reap the benefits of recycling by assisting with research, policy initiatives, coalition building, and technical assistance.

PUBLICATIONS: The institute publishes comprehensive studies based on applied research and policy analysis (a pamphlet listing the various publications is available through the ILSR office). Specific recycling literature, including a poster and various monographs, can be found through ILSR's World Wide Web page.

Institute of Scrap Recycling Industries, Inc. (ISRI)
1325 G Street, NW, Suite 1000
Washington, DC 20005-3104
(202) 737-1770
(202) 626-0900 (FAX)
WWW: http://www.isri.org
Jim Fisher, President

ISRI—The Original Recyclers®—is a trade association representing approximately 1,600 companies that process, broker, and consume scrap commodities. Such commodities include ferrous and nonferrous metals, paper, glass, textiles, rubber, and plastic. The membership also includes suppliers of equipment and services that serve the scrap industry. Founded in 1987, ISRI was formed by the merger of the National Association of Recycling Industries (founded in 1913) and the Institute of Scrap Iron and Steel (founded in 1928). Each year, ISRI members handle an average of 90 to 100 million-plus tons of recyclables destined for domestic use and overseas markets. ISRI members thus consider themselves to be experts in the field of recycling and resource recovery. They offer this expertise to communities and organizations in planning, establishing, and implementing recycling activities. A primary objective of ISRI is to "bring about a greater awareness of the industry's role in conserving the future through recycling" and in increasing recycling opportunities by promoting the concept of designing products with eventual recycling in mind (the Design for Recycling® program).

PUBLICATIONS: ISRI publishes a quarterly public relations magazine, *Phoenix: Voice of the Scrap Recycling Industries,* and offers subscriptions to *SCRAP Processing and Recycling* magazine. The organization also offers a publication called *Value,* as well as a fact sheet on the industry. A catalog of additional publications and audio visuals is also available from ISRI.

Keep America Beautiful, Inc. (KAB)
1010 Washington Boulevard
Stamford, CT 06901
(203) 323-8987
(203) 325-9199 (FAX)
Roger W. Powers, President

KAB was founded in 1953 specifically to confront the issue of litter, and it developed a network of community-based programs to handle this problem. Today, KAB affiliates are adapting their programs to "encourage recycling and educate the public about solid waste disposal" in order to improve waste handling practices in U.S. communities.

PUBLICATIONS: KAB publishes several informational pamphlets as well as its monthly award-winning newsletter, *Network,* and specializes in providing information on organizing community cleanup projects. An extensive study, *The Role of Recycling in Integrated Solid Waste Management to the Year 2000,* is available through KAB, and its Solid Waste Task Force publishes a bimonthly newsletter called *FOCUS: Facts on Municipal Solid Waste.* In addition, KAB publishes a catalog containing educational and promotional materials, including videotapes, posters, and buttons.

National Association for Plastic Container Recovery (NAPCOR)
100 N. Tryon Street, Suite 3770
Charlotte, NC 28202
(704) 358-8882
(704) 358-8769 (FAX)
Toll-free PET plastic information hotline: 1-800-7NAPCOR
WWW: http://www.napcor.com
E-mail: NAPCOR@aol.com
Quinn Davidson, President

NAPCOR is a nonprofit trade association formed in 1987 with the objective of facilitating the economic recovery of plastic bottles, with an emphasis on PET plastic, "through collection, reclamation, and development of end uses for postconsumer bottles." NAPCOR also promotes the "environmental efficiency of PET plastic bottles" and the end products developed from these bottles.

PUBLICATIONS: NAPCOR produces a variety of advertising materials about recycling PET. Several brochures, "Remove, Rinse, Recycle," "Flat's Where It's At," and "Look Out for Number 1!" are also available, along with a purchasing guide for recycled PET plastic products. *PET Projects* is NAPCOR's newsletter.

National Association of Towns and Townships (NATaT)
444 N. Capitol Street, NW, Suite 294
Washington, DC 20001
(202) 624-3550
(202) 624-3554 (FAX)
Tom Halicki, Director

NATaT is the only national organization exclusively devoted to smaller hometown governments in the United States. Operating as a nonprofit, its purpose is to "strengthen the effectiveness of towns, townships, and small communities in the United States and to promote their interests in the public and private sectors." One of the resources offered by NATaT to its members is a recycling program geared toward small towns: "Why Waste a Second Chance?"

PUBLICATIONS: Although the statistics in "Why Waste a Second Chance?" have not been updated since it first came out in 1989, the program remains a good introduction to recycling for a small community, and includes a guidebook, a VHS video, and a video user's guide. The videotape may be rented or bought; guidebooks may be ordered in quantity. NATaT also produces a monthly newsletter, the *Washington Report*, containing legislative updates, resource tips, and small community news.

National Consumers League (NCL)
1701 K Street, NW, Suite 1200
Washington, DC 20006
(202) 835-3323
(202) 835-0747 (FAX)
Linda F. Golodner, President

NCL is a nonprofit, membership organization representing consumers and workers on the federal level since 1899. NCL provides consumers with useful information on issues surrounding recycling and educates the public on topics ranging from energy conservation to child labor.

PUBLICATIONS: NCL publishes a bimonthly newsletter, the *NCL BULLETIN*, and a brochure specifically dealing with solid waste and recycling, *The Garbage Problem: Effective Solutions for Consumers*. NCL also produces several *Green Sheets* on recycling and waste reduction resources.

National Recycling Coalition (NRC)
1727 King Street, Suite 105
Alexandria, VA 22314-2720
(703) 683-9025 x402
(703) 683-9026 (FAX)
Will Ferretti, Executive Director

Founded in 1978, NRC is the largest organization in the country that deals exclusively with recycling issues. Its board of directors and committee chairpersons include many leaders from other environmental organizations, as well as corporate executives, municipal administrators, and other professionals in the field of recycling. NRC provides a national forum to promote recycling through a range of integrated programs, including professional education and networking, policy development, market development, advocacy, source reduction, research and analysis, community outreach and leadership, recovered material supply, business leadership in recycling, recycled materials manufacturing technology assistance, and media outreach. NRC has also created the Buy Recycled Business Alliance, which promotes the buying of recycled-content products in its member companies.

PUBLICATIONS: NRC publishes a quarterly newsletter, *The NRC Connection*, and its members may receive discounts on subscriptions to the following recycling periodicals: *Resource Recycling*, *BioCycle*, *Waste Age*, and *Recycling Times*.

National Soft Drink Association (NSDA)
1101 16th Street, NW
Washington, DC 20036
(202) 463-6700 (Solid Waste Programs Division)
(202) 659-5349 (FAX)
WWW: http://www.nsda.org
E-mail: mariec@nsda.com
Gifford Stack, Vice President of NSDA and Solid Waste Programs Director

NSDA is a national trade association representing the entire American soft drink industry, which encompasses approximately 560 soft drink bottling firms in the United States. The association's members include companies that manufacture and market soft drinks, suppliers to the industry, and franchise companies. In April 1989, NSDA created the Solid Waste Programs Division in order to better articulate the industry's proactive commitment to recycling and comprehensive solid waste management. This division considers its mission "to recycle more of our containers." To this end, it provides hands-on assistance to bottler members; works closely with state soft drink associations, elected officials, and administrators; and consults with professional recyclers, opinion leaders, and other trade associations in the recycling field.

PUBLICATIONS: The Solid Waste Programs Division currently publishes a quarterly newsletter entitled *The Soft Drink RECYCLER* and is in the process of producing a number of new publications on the subjects of recycling and environmentally conscientious packaging as this book goes to press. NSDA's World Wide Web page, launched in 1997, will keep current on solid waste issues.

National Solid Wastes Management Association (NSWMA)
4301 Connecticut Avenue, NW, Suite 300
Washington, DC 20008
(202) 244-4700
(202) 966-4818 (FAX)
Toll-free hotline: 1-800-424-2869
Sheila Prindiville, Executive Vice-President

NSWMA is a national trade association representing 2,500 member companies that are involved in private waste collection and disposal, recycling, landfill and waste-to-energy facility design and operation, and medical and hazardous waste treatment and disposal. Following a restructuring in December 1993, NSWMA became part of the Environmental Industry Associations (see listing above). The association also has two nonactive membership categories for individuals or groups with an interest in receiving information from NSWMA. Associate memberships are offered to individuals who are not in the waste business, and affiliate memberships are available to public officials, nonprofit organizations, and government agencies. Several professional institutes and councils are managed by NSWMA, including the Waste Recyclers Council (WRC), which works to "promote recycling efforts within and beyond the industry, and focus greater attention on recyclable materials markets."

PUBLICATIONS: NSWMA produces an extensive list of periodicals, brochures, indexes, fact sheets, reprints, and special reports, many of which deal specifically with recycling (see Chapter 7).

North American Association for Environmental Education (NAAEE)
P.O. Box 400
Troy, OH 45373-0400
(937) 676-2514 (phone and FAX)
Edward McCrea, Executive Director

Founded in 1971, NAAEE is dedicated to promoting environmental education and supporting its educators around the world. The Environmental Issues Forum, one of the programs run by NAAEE, includes a section on solid waste problems and solutions.

PUBLICATIONS: *The Solid Waste Mess: What Should We Do with the Garbage?* is one of NAAEE's issue books, which is accompanied by an *Issues in Brief* booklet. The *Environmental Communicator* is a bimonthly newsletter published by NAAEE. A list of additional publications is also available.

Plastic Bag Association (PBA)
Plastic Bag Information Clearinghouse
1817 E. Carson Street
Pittsburgh, PA 15203
(412) 381-8890 (FAX)
355 Lexington Avenue, 17th Floor
New York, New York 10017
(212) 661-4261
Toll-free plastic bag recycling information: 1-800-438-5856
WWW: http://www.plasticbag.com/pba.html
E-mail: pbainfo@aol.com
Holly Munter, Executive Director

With offices in both New York and Pittsburgh, PBA is a consortium of 60 plastic bag manufacturers and suppliers who work together to promote the industry as well as provide members with programs, services, and the forum for addressing environmental, regulatory, and other plastic bag issues.

PUBLICATIONS: The Plastic Bag Information Clearinghouse is a resource for "consumers, media, and others to obtain up-to-date environmental information about plastic bags" and is accessible by the toll-free number given above or by e-mail. The brochure, "The Life of a Plastic Bag" summarizes how plastic bags fit into the 3Rs, and the Clearinghouse also maintains a database of over 14,000 plastic bag recycling sites across the United States.

Plastic Lumber Trade Association (PLTA)
540 S. Main Street, Building 7
Akron, OH 44311-1023
(216) 762-1963
Toll-free information hotline: 1-800-886-8990
Alan Robbins, President

PLTA's goal is to unite "diverse plastic lumber producers to promote the engineering, art and science, and the marketing and procurement of recycled plastic lumber and other recycled plastic lumber products." PLTA is working on standardizing testing procedures, promoting standards of quality, encouraging cooperation within the industry, and promoting construction products made from recycled plastics.

Plastics and Composites Group
Rutgers, the State University of New Jersey
Livingston Campus, Building 4109
New Brunswick, NJ 08903
(908) 445-3632 and 445-5236
(908) 445-5636 (FAX)
Tom Nosker, Director

The Center for Plastics Recycling Research (CPRR) was founded and funded by the Plastics Recycling Foundation (PRF) in the 1980s to quantify facts about recycling plastics. As of 1 October 1996, the CPRR was closed at Rutgers and replaced by a smaller research organization, the Plastics and Composites Group (PCG). The PCG does research in the fields of civil and environmental engineering, with a large focus on the plastic lumber industry and a smaller focus on resin recovery from recycling programs.

Polystyrene Packaging Council (PSPC)
1801 K Street, NW, Suite 600
Washington, DC 20006
(202) 974-5321
(202) 296-7354 (FAX)
WWW: http://www.polystyrene.org
Michael Levy, Executive Director

PSPC was founded in 1988 by about 35 polystyrene manufacturers as a special purpose group of the Society of the Plastics Industry, Inc. This organization works with "local governments and the public to increase awareness of polystyrene recycling, answer questions about polystyrene manufacture and use, and help those who want to collect polystyrene find a nearby recycler."

PUBLICATIONS: Several fact sheets on polystyrene recycling and the environment are published by PSPC for free distribution to the public. In addition, a video entitled *Solid Waste Solutions* is also available for educational efforts. PSPC's World Wide Web page also offers a wealth of information on recycling polystyrene.

Renew America
1400 16th Street, NW, Suite 710
Washington, DC 20036
(202) 232-2252
(202) 232-2617 (FAX)
Debbie Sliter, Executive Director

Renew America is a nonprofit, membership-funded organization that works to "provide others with examples of environmental programs that work." Its mission is "to spread the word about working solutions to today's environmental problems so that people can more easily improve their communities."

PUBLICATIONS: The organization publishes the *Environmental Success Index,* which is a clearinghouse of verified, working programs that protect, restore, and enhance the environment. The index is also part of the group's Searching for Success Program, which attempts to identify, promote, and reward successful environmental programs across the nation. One of the 20 issues highlighted in this program is solid waste reduction and recycling.

Society of the Plastics Industry, Inc. (SPI)
1801 K Street, NW, Suite 600K
Washington, DC 20006-1301
(202) 974-5200
(202) 296-7005 (FAX)
Larry L. Thomas, President

SPI is the principal trade association for the U.S. plastics industry, with 2,000 members representing all aspects of that industry. Founded in 1937, it is a well-established organization including dozens of divisions, market councils, and special purpose groups devoted to every aspect of plastics use and manufacture. SPI sponsors conferences throughout the United States and maintains regional offices in Washington, D.C.; Des Plaines, Illinois; Weymouth, Massachusetts; and Long Beach, California.

PUBLICATIONS: SPI's literature catalog includes several hundred publications on plastics, ranging from public information brochures to technical research papers.

Solid Waste Association of North America (SWANA)
P.O. Box 7219
Silver Spring, MD 20907-7219
(301) 585-2898
(301) 589-7068 (FAX)
Dr. John Skinner, Executive Director

Founded in 1961, SWANA is a nonprofit educational organization that serves individuals and communities responsible for the management and operation of municipal solid waste management systems. Its membership includes professionals from both the public and private sectors, and it is the largest member-based solid waste management association in the world. "Dedicated to the advancement of professionalism in the field," SWANA offers several certified training courses, technical assistance, and other educational opportunities. It has chapters throughout the United States.

PUBLICATIONS: SWANA publishes a monthly newsletter, *MSW Solutions*, and produces a catalog of mostly technical publications.

Southwest Public Recycling Association (SPRA)
P.O. Box 27210
Tucson, AZ 85726-7210
(520) 791-4069
(520) 791-5242 (FAX)
Gary J. Olson, Executive Director

SPRA is a regional effort of over 20 municipalities in five states. Its goals are to: "encourage waste reduction and reuse; increase recyclables recovered from the waste stream; increase utilization of recycled and recovered products; promote manufacturing processes that use recycled materials; and promote the buy recycled philosophy." The nonprofit association was founded in January 1991 by mayors and staff from municipalities in Arizona, Colorado, Nevada, New Mexico, Utah, and west Texas. Currently, SPRA represents over 5 million people from 104 political jurisdictions.

PUBLICATIONS: The *SPRA News* is a newsletter that is published on a quarterly basis. A list of publications, mostly technical reports and fact sheets, is also available from SPRA.

Steel Recycling Institute (SRI)
680 Andersen Drive
Foster Plaza 10
Pittsburgh, PA 15220-2700
(412) 922-2772
(412) 922-3213 (FAX)
Toll-free consumer information hotline: 1-800-937-1226
William H. Heenan, President

SRI is a nonprofit industry-sponsored association with the mission of promoting and sustaining steel can recycling throughout the United States and serving as an "information and technical resource to those who are interested in recycling steel." This association is available to help community leaders, recyclers, business people, and educators develop steel can recycling programs and promote them to the public. SRI has established seven regional offices in addition to its main headquarters in Pittsburgh, and maintains its national toll-free number to answer questions and provide technical assistance to consumers interested in recycling steel products from cans to washing machines.

PUBLICATIONS: SRI publishes a quarterly newsletter, *The Recycling Magnet,* that is devoted to publishing interesting stories about steel can recycling. Copies of several videotapes are available through SRI for previewing by government officials, recycling coordinators, and educators. Additionally, SRI has developed several educational programs for all ages, preschool to adult, as well as a variety of fact sheets and a poster, which are available for purchase. A number of informational brochures are also produced by SRI on how to recycle specific steel products (e.g., food cans, aerosol cans, paint cans, appliances, and automobiles).

U.S. Environmental Protection Agency (EPA)
Office of Solid Waste
401 M Street, SW
Mail Code: 5305W
Washington, DC 20460
(703) 308-8895
Hotline for recycling brochures: 1-800-424-9346
WWW: http://www.epa.gov
Michael H. Shapiro, Director, Office of Solid Waste

EPA is the federal agency that deals with solid waste and recycling, as well as all other environmental issues. It is also responsible for administering the Resource Conservation and Recovery Act (RCRA).

PUBLICATIONS: The Environmental Protection Agency has an extensive list of educational publications on solid waste and recycling, which can be ordered by calling the EPA hotline. In addition, the Office of Solid Waste publishes a quarterly newsletter on recycling entitled *Reusable News,* which is available to the public at no cost (see Chapter 7 for more information).

EPA also has ten regional offices throughout the United States, each of which coordinates its own recycling education programs and produces additional recycling information specific to the area it administers. The names and phone numbers listed below refer to primary contacts in each regional division's Office of Solid Waste:

U.S. EPA Region 1
90 Canal Street
Mail Code: HER-CAN6
Boston, MA 02203
(617) 573-9678
(617) 573-9662 (FAX)
Chuck Franks

Region 1: Connecticut, Maine, Massachusetts, New Hampshire, Rhode Island, and Vermont.

U.S. EPA Region 2
290 Broadway
Mail Code: 2AWM
New York, NY 10007-1866
(212) 637-4100
(212) 637-4437
Stanley Siegel

Region 2: New Jersey, New York, Puerto Rico, and the Virgin Islands.

U.S. EPA Region 3
841 Chestnut Street
Mail Code: 3HW60
Philadelphia, PA 19107
(215) 566-3374
(215) 566-3114 (FAX)
Jeff Alper

Region 3: Delaware, Washington, DC, Maryland, Pennsylvania, Virginia, and West Virginia.

U.S. EPA Region 4
345 Courtland Street, NE
Mail Code: 4WD-OSW
Atlanta, GA 30365
(404) 562-4300
(404) 347-5205 (FAX)
Kelly Ewing

Region 4: Alabama, Florida, Georgia, Kentucky, Mississippi, North Carolina, South Carolina, and Tennessee.

U.S. EPA Region 5
77 W. Jackson Boulevard
Mail Code: DRP-8J
Chicago, IL 60604-3590
(312) 886-0976
(312) 353-4788 (FAX)
Mary Setnicar
Region 5: Illinois, Indiana, Michigan, Minnesota, Ohio, and Wisconsin.

U.S. EPA Region 6
First Interstate Bank Tower
1445 Ross Avenue
Mail Code: 6H-H
Dallas, TX 75202-2733
(214) 665-6707
(214) 665-6762 (FAX)
Paul Thomas

Region 6: Arkansas, Louisiana, New Mexico, Oklahoma, and Texas.

U.S. EPA Region 7
726 Minnesota Avenue
Mail Code: ARTX/TSPP
Kansas City, KS 66101
(913) 551-7523
(913) 551-7947 (FAX)
David Flora

Region 7: Iowa, Kansas, Missouri, and Nebraska.

U.S. EPA Region 8
999 18th Street, Suite 500
Mail Code: 8P2-P2
Denver, CO 80202-2466
(303) 312-6356
(303) 312-6191 (FAX)
Judy Wong

Region 8: Colorado, Montana, North Dakota, South Dakota, Utah, and Wyoming.

U.S. EPA Region 9
75 Hawthorne Street
Mail Code: H-W-3
San Francisco, CA 94105
(415) 744-2106
(415) 744-1044 (FAX)
David Duncan

Region 9: Arizona, California, Hawaii, Nevada, Guam, and American Samoa.

U.S. EPA Region 10
1200 Sixth Avenue
Mail Code: HW-107
Seattle, WA 98101
(206) 553-1716
(206) 553-8509 (FAX)
Dave Croxton

Region 10: Alaska, Idaho, Oregon, and Washington.

Worldwatch Institute
1776 Massachusetts Avenue, NW
Washington, DC 20036
(202) 452-1999
(202) 296-7365 (FAX)
Hotline for publications/subscription information: 1-800-555-2028
Lester R. Brown, President

"The Worldwatch Institute is an independent, nonprofit research organization created to analyze and to focus on global problems." The group considers that solid waste is one of those problems and believes recycling and reuse are among the solutions.

PUBLICATIONS: The institute publishes a series of Worldwatch Papers, several of which deal specifically with recycling and solid waste issues. The bimonthly magazine *World Watch,* an annual State of the World report, and the Worldwatch Environmental Alert book series are also published by the institute.

Reference Materials

In a search through the available literature, the subject of recycling can be found by itself, within texts that discuss environmental issues, and in studies on solid waste maagement. This selected bibliography provides a wide perspective of books, periodicals, curriculum guides, government documents, and popular articles, as well as videocassettes and a computer software program.

Books, Brochures, Reports, and Directories

Recycling, Composting, Solid Waste, and General Environmental Issues

Ackerman, Frank. **Why Do We Recycle?: Markets, Values, and Public Policy.** Washington, DC: Island Press, 1996. 180p. $16.95. ISBN 1-55963-505-3.

A somewhat controversial study on recycling programs, with a thoughtful assessment of environmental ethics versus common economic policies. The author examines the connection between recycling and market values. Ackerman points out the value of sustainable economics and the necessity of recycling in such a system.

Appelhof, Mary. **Worms Eat My Garbage: How to Set Up and Maintain a Worm Composting System.** Kalamazoo, MI: Flower Press, 1997. $10.95. ISBN 0-942256-10-7.

This is the most current edition of a series of books written by Appelhof on this subject. She is widely recognized as the foremost expert on home vermi-composting. This book will tell you everything you've ever wanted to know about composting food wastes in your home via red worms.

Aquino, John T., ed. **Waste Age/Recycling Times' Recycling Handbook.** Boca Raton, FL: Lewis Publishers, 1995. 281p. $59.95hc. ISBN 1-56670-068-X.

A comprehensive look at recycling from a more technical standpoint, from the editors of *Waste Age* and *Recycling Times* magazines.

Asher, Marty, and, Tom Christopher. **Compost This Book! The Art of Composting for Your Yard, Your Community, and the Planet.** San Francisco, CA: Sierra Club Books, 1994. 248p. $12. ISBN 0-87156-596-X.

A very entertaining, and informative, introduction to the art and science of composting. This book covers everything from historical anecdotes to cricket manure.

Baldwin, J., ed. **Whole Earth Ecolog.** New York: Harmony Books, 1990. 129p. $15.95. ISBN 0-517-57658-9.

A catalog containing "the best of environmental tools and ideas." This collection of products and resources is arranged by subjects such as recycling, urban ore, and composting. Information on other publications and environmental essays is also included.

BioCycle Staff, eds. **The BioCycle Guide to Collecting, Processing and Marketing Recyclables.** Emmaus, PA: BioCycle, 1991. 229p. $49.95. ISBN 0-932424-11-2.

An all-in-one reference on successful recycling for community and industry, published by *BioCycle, Journal of Waste Recycling.* This volume contains comprehensive data on developing curbside collection programs; systems and equipment analysis, together with advice on marketing materials; and a special report on MRFs. It is a valuable guide for all involved in waste reduction, recycling, resource conservation, and general environmental improvement.

———. **The BioCycle Guide to Yard Waste Composting.** Emmaus, PA: BioCycle, 1991. 197p. $49.95. ISBN 0-932424-10-4.

A very complete reference guide to starting and operating a community yard waste composting program efficiently. From the editors of *BioCycle* magazine, which has been reporting on scientific studies of composting since 1960, this book includes sections on planning, collection, site selection and regulations, methods and operations, and composting principles. A directory of equipment suppliers, a section on marketing and utilization, and cost and economic comparisons are also included.

Blumberg, Louis, and Robert Gottlieb. **War on Waste: Can America Win Its Battle with Garbage?** Washington, DC: Island Press, 1989. 325p. $34.95. ISBN 0-933280-91-0.

This book, written for the average person who is not an expert in the field of solid waste, provides an in-depth analysis of the fundamental causes of our current garbage crisis. It offers suggestions and proposals for public officials, community organizations, and concerned citizens who must find solutions to solid waste problems in their locales.

Branson, Gary D. **The Complete Guide to Recycling at Home.** Charlottesville, VA: Betterway Publications, 1991. 150p. Appendixes. $14.95. ISBN 1-55870-189-3.

Emphasizes source reduction before recycling, with suggestions for changes in the home. This volume also includes information on energy conservation, and an appendix lists sources of products for use in recycling programs as well as those made from recycled materials.

Campbell, Stu. **Let It Rot! The Gardener's Guide to Composting.** Pownal, VT: Storey Communications, 1990. 152p. $8.95. ISBN 0-88266-635-5.

First published in 1975, this updated edition covers all one needs to know about household and backyard composting. It is written in an easy-to-grasp manner and covers everything from choosing the right materials to compost to buying the right equipment for your needs.

Chandler, William U. **Worldwatch Paper 56, Materials Recycling: The Virtue of Necessity.** Washington, DC: Worldwatch Institute, 1983. 52p. $4. ISBN 0-916468-55-0.

Gives a good, concise overview to the solid waste problems facing the world. Although the statistics are outdated, the basic principles and statements presented here are still current. The author also addresses the economic requirements that societies must establish to make recycling work—that consumers must pay the true environmental costs of the products they use, that world markets for scrap must be built and developed, and that collection of recyclable wastes must increase.

Conway, William E. **Winning the War on Waste.** Nashua, NH: Conway Quality, 1996. 250p. $6.95. ISBN 0-9631464-4-0.

This book, geared toward businesses, communities, and government organizations, provides a practical approach to waste reduction through multifaceted solutions. Conway explains how to increase quality and decrease waste.

Denison, Richard A., and John Ruston. **Recycling and Incineration: Evaluating the Choices.** Covelo, CA: Island Press, 1990. 322p. $34.95hc. ISBN 1-55963-055-8.

Examines the technology, economics, environmental concerns, and legal intricacies behind recycling and incineration as two high-priced approaches to tackling a community's solid waste problems. Rather than looking at two parts of an integrated solid waste system, the authors treat the issues in an "either/or" fashion. A good overview of the basics of waste reduction, recycling, and incineration, this book offers cost comparisons of the two approaches, an evaluation of the perceived health and environmental impacts, and a road map through the tangle of regulations. This is a book for policymakers, waste management officials, municipal employees, and concerned citizens who are analyzing different solutions to solid waste problems.

The EarthWorks Group, eds. **50 Simple Things You Can Do to Save the Earth.** Berkeley, CA: EarthWorks Press, 1989. 96p. $4.95. ISBN 0-929634-06-3.

A concise list of practical suggestions that any individual or family can use to tackle the environmental problems we all face. Many of the ideas relate to various aspects of recycling, and further resources are given.

———. **50 Simple Things Your Business Can Do to Save the Earth.** Kansas City, MO: Andrews and McMeel, 1991. 120p. $6.95. ISBN 1-879682-02-8.

More projects and ideas from EarthWorks, specifically geared toward businesses. Office recycling programs, source reduction, and waste exchanges between industries are among the solid waste solutions addressed in this book.

———. **The Next Step: 50 More Things You Can Do to Save the Earth.** Kansas City, MO: Andrews and McMeel, 1991. 120p. $5.95. ISBN 0-8362-2302-0.

An overview of various successful environmental projects throughout the United States and Canada. This is essentially a reference book on how to organize and educate one person or an entire community on specific environmental issues. Several of the projects deal with recycling and issues surrounding solid waste.

———. **The Recycler's Handbook.** Berkeley, CA: EarthWorks Press, 1990. 132p. $4.95. ISBN 0-929634-08-X.

Another well-composed how-to book by the EarthWorks Group. This volume deals specifically with all aspects of recycling in the home, school, or office, and includes tips for starting community recycling programs.

Goldbeck, David, and Nikki Goldbeck. **Choose to Reuse: An Encyclopedia of Services, Businesses, Tools, and Charitable Programs That Facilitate Reuse.** Woodstock, NY: Ceres Press, 1995. 455p. $15.95. ISBN 0-9606138-6-2.

This book provides more than 2,000 resources for reusing a vast variety of materials that might ordinarily be tossed in the trash (over 200 items are listed, from Air Filters to Zippers). "Choice Stories" about many of the topics are entertaining as well as educational.

Graham, Kevin, and Gary Chandler. **Environmental Heroes—Success Stories of People at Work for the Earth.** Boulder, CO: Pruett Publishing, 1996. 244p. $16.95. ISBN 0-87108-866-5.

This readable and informative book includes profiles of people involved in a variety of recycling and composting projects, as well as numerous other environmental issues. Includes resources for gathering follow-up material on any of the subjects mentioned.

Harmonious Technologies. **Backyard Composting: Your Complete Guide to Recycling Yard Clippings.** Ojai, CA: Harmonious Technologies, 1995. 96p. $6.95. ISBN 0-9629768-3-0.

Filled with helpful illustrations, this is a good step-by-step guide for the beginning backyard composter. Included are instructions on how to build your own compost bins, how to maintain "healthy" compost, and suggestions for what to do with the finished product.

Hassol, Susan, and Beth Richman. **Recycling: 101 Practical Tips for Home and Work.** Snowmass, CO: Windstar Foundation, 1989. 91p. $3.95. ISBN 0-9622492-3-8.

A handbook in the Windstar EarthPulse series providing practical tips on how to reduce, reuse, and recycle in the home and at work. Background information on the 3Rs, landfills, and other waste management measures, interesting statistics on solid waste recycling, and resources for recycled products and products to be used in recycling programs are also included. One section offers recycling suggestions and projects for paper, glass, aluminum, scrap, plastic, organic material, and household hazardous waste.

Hershkowitz, Dr. Allen, and Dr. Eugene Salerni. **Garbage Management in Japan—Leading the Way.** New York: INFORM, 1987. 152p. $15. ISBN 0-918780-43-8.

A comprehensive report on Japan's sophisticated solid waste management techniques, including recycling. Documenting Japan's evolution in the garbage business, this report offers American planners and investors a lesson plan to help them avoid economic pain and environmental degradation. Japan has one of the highest recycling rates in the world, primarily because the country has so many people and so little space for landfills.

Hollender, Jeffrey. **How to Make the World a Better Place.** New York: William Morrow, 1990. 303p. $9.95. ISBN 0-688-08479-6.

A general book about global environmental issues, with an informative chapter on waste and recycling. Littering and source reduction are also addressed.

INFORM. **Business Recycling Manual.** Washington, DC: INFORM, 1991. 196p. $92 (binder format). ISBN 0-918780-57-8.

A complete guide to business recycling systems for recycling managers. This manual begins with the program development phase, explaining how to gather data and conduct a waste audit, and includes information on monitoring and evaluating different recycling programs and marketing recyclables. Worksheets are included.

Jones, Teresa, Edward J. Calabrese, Charles E. Gilbert, and Alvin E. Winder. **Solid Waste Education Recycling Directory.** Chelsea, MN: Lewis Publishers, 1990. 109p. $57hc. ISBN 0-87371-359-1.

A state-by-state guide to recycling education curricula for kindergarten through twelfth grade. This directory provides information on ordering the programs and itemizes the costs of materials.

Lamb, Marjorie. **2 Minutes a Day for a Greener Planet.** San Francisco, CA: Harper & Row, 1990. 243p. $7.95. ISBN 0-06-250507-6.

Informative, easy-to-read suggestions for improving the environment one step at a time. Although several chapters address solid waste issues with regard to saving paper, excess packaging, reuse, and recycling, this is basically another broad look at multiple issues, with little in-depth treatment.

Lund, Herbert F. **The McGraw-Hill Recycling Handbook.** New York, NY: McGraw-Hill, 1992. $87.50hc. ISBN 0-07039096-7.

This reference book on recycling is geared toward officials who are responsible for creating and implementing recycling programs in their cities, businesses, or schools. It is especially helpful for providing information on cost analysis for such programs, and includes information on facilities designs and recycling equipment.

Martin, Deborah L., and Grace Gershuny, eds. **The Rodale Book of Composting.** Emmaus, PA: Rodale Press, 1992. 278p. $14.95. ISBN 0-87857-991-5.

This is one of the most comprehensive books written on the subject of composting, with simple instructions, ideas for troubleshooting, and lots of illustrations. Vermicomposting (using red worms) and indoor composting during the winter are also covered in this thorough guide.

National Solid Wastes Management Association. **Garbage Then and Now.** Washington, DC: National Solid Wastes Management Association, 1990. 8p. $1.50. Order number: 003118.

A garbage timeline from 500 B.C. to the 1990s. Fun and informative, this is a good educational tool for children and adults alike and can be ordered in packages of 50 for $18.75.

———. **Public Attitudes Toward Garbage Disposal.** Washington, DC: National Solid Wastes Management Association, 1990. 11p. $7.50. Order number: 910783

Examines public attitudes about recycling, incineration, and landfilling. This volume also surveys state recycling goals across America.

———. **Recycling Solid Waste at-a-Glance.** Washington, DC: National Solid Wastes Management Association, 1991. 4p. $1.50. Order number: 906419.

A quick-reference fact sheet on recycling, including product-by-product rates and information on what is involved in recycling, what makes it work, and what impedes its progress.

———. **Resource Recovery in the United States.** Washington, DC: National Solid Wastes Management Association, 1989. 8p. $7.50. Order number: 906446.

A statistical analysis, with charts and tables, of recycling and waste-to-energy programs and how they integrate into comprehensive solid waste planning. This work addresses the problems associated with waste-to-energy plants in an easy-to-read fashion.

———. **Waste Recycler's Council Directory.** Washington, DC: National Solid Wastes Management Association, 1989. 36p. $5. Order number: 908624.

A comprehensive list of members of the Waste Recycler's Council. In 1989, council members handled nine million tons of recyclable materials and spent over $385 million for facilities and equipment to collect and process those recyclables.

Platt, Brenda. **Beyond 40 Percent: Record-Setting Recycling and Composting Programs.** Covelo, CA: Island Press, 1991. 270p. $27.50. ISBN 1-55963-073-6.

A practical guide for communities seeking solutions to solid waste problems, this book is produced by the Institute for Local Self-Reliance. Based on case studies of17 cities in the United States, it includes information on collection and processing methods, equipment, and costs. A local contact is given for each featured community to help the reader communicate with officials who have established successful recycling programs.

Pollock, Cynthia. **Worldwatch Paper 76, Mining Urban Wastes: The Potential for Recycling.** Washington, DC: Worldwatch Institute, 1987. 58p. $4. ISBN 0-916468-77-1.

A Worldwatch Paper focusing on urban solid waste that could potentially be recycled. Pollock promotes the environmental benefits of using secondary, or recycled, materials instead of virgin resources and addresses the issues of household hazardous wastes entering landfills, as well as food and yard waste composting. Although the statistics are not current, the basic precepts and conclusions presented in this report still hold true for today's urban areas.

Resource Recycling. **The Directory of On-line Resources in Recycling and Composting.** Portland, OR: Resource Recycling, 1996. 26p. $12.95. Order number: OL1.

Compiled by the editors of *Resource Recycling* magazine, this directory lists World Wide Web pages, bulletin board systems dealing with solid waste management, and hundreds of e-mail addresses. Databases, government resources, and online mailing lists are also included.

Rifkin, Jeremy, ed. **The Green Lifestyle Handbook—1001 Ways You Can Heal the Earth.** New York: Henry Holt, 1990. 198p. $10.95. ISBN 0-8050-1369-5.

A book about global environmental problems, with solutions that individuals and communities can implement. One chapter deals specifically with recycling and solid waste.

Steel Can Recycling Institute. **Buy Recycled with Recyclable Steel.** Pittsburgh, PA: Steel Recycling Institute, 1996. 6p. Free.

———. **Recycling Empty Steel Aerosol Cans in Curbside and Drop-Off Programs.** Pittsburgh, PA: Steel Recycling Institute, 1996. 6p. Free.

———. **Recycling Scrapped Automobiles.** Pittsburgh, PA: Steel Recycling Institute, 1996. 6p. Free.

———. **Recycling Steel Appliances.** Pittsburgh, PA: Steel Recycling Institute, 1996. 6p. Free.

———. **Recycling Steel Cans From Food Service Facilities.** Pittsburgh, PA: Steel Recycling Institute, 1996. 6p. Free.

———. **Residential Recycling of Empty Steel Paint Cans.** Pittsburgh, PA: Steel Recycling Institute, 1996. 6p. Free.

———. **Steel Recycling Starts in the Home.** Pittsburgh, PA: Steel Recycling Institute, 1996. 6p. Free.

These brochures on the various aspects of steel recycling are aimed at consumers and are filled with practical and educational information. All these publications are available through the Steel Recycling Institute (SRI) and can be ordered by calling 1-800-937-1226, x289, or by writing SRI at 680 Andersen Drive, Pittsburgh, PA 15220-2700 (see Chapter 6).

Thompson, Claudia G. **Recycled Papers: The Essential Guide.** Cambridge, MA: MIT Press, 1992. 162p. $25. ISBN 0-262-70046-8.

A clear and concise layman's guide to recycled printing and writing papers. Thompson is a graphic designer who has researched recycled papers since 1988. The dimensions of the solid waste problem, a history of papermaking, and an examination of the varieties and characteristics of available papers are included in this text, which is printed on four different types of recycled paper.

Underwood, Joanna D., Dr. Allen Hershkowitz, and Dr. Maarten de Kadt. **Garbage: Practices, Problems and Remedies.** New York: INFORM, 1988. 32p. $3.50. ISBN 0-918780-47-0.

A source of concise, accurate data on solid waste management practices in the United States. Presenting useful facts about the garbage Americans create and how they manage it, this booklet also identifies waste management alternatives such as source reduction, recycling, improved incineration, and better planned and monitored landfills.

U.S. Environmental Protection Agency. **A Catalogue of Hazardous and Solid Waste Publications.** 5th ed. Washington, DC: EPA, 1991. 61p. Free. EPA/530-SW-91-013.

A catalog listing the most-requested documents of the Office of Solid Waste. Many of the documents are highly technical, but a few on recycling are geared to the general public (and therefore are included in this reference section). A number of documents about the Resource Conservation and Recovery Act (RCRA) are also included.

———. **Characterization of MSW in the U.S.: 1995 Update, Executive Summary.** Washington, DC: EPA, 1996. 13p. Free. EPA/530-S-96-001.

This booklet summarizes the findings of the EPA's standard reference for waste composition, generation, and management numbers. This report is prepared each year by Franklin Associates, using the EPA's flow methodology.

———. **How to Set Up a Local Program to Recycle Used Oil.** Washington, DC: EPA, 1989. 41p. Free. EPA/SW-89-039a.

———. **Recycling Used Oil: For Service Stations and Other Vehicle-Service Facilities.** Washington, DC: EPA, 1989. Free. EPA/530-SW-89-03d.

———. **Recycling Used Oil: 10 Steps to Change Your Oil.** Washington, DC: EPA, 1989. Free. EPA/SW-89-039c.

———. **Recycling Used Oil: What Can You Do?** Washington, DC: EPA, 1989. 4p. Free. EPA/530-SW-89-039b.

The four booklets listed above provide most of the information necessary to recycle used oil or set up a program to do so. These booklets are all available through the RCRA/Superfund hotline: 1-800-424-9346.

———. **Recycle.** Washington, DC: EPA, 1988. 4p. Free. EPA/530-SW-88-050.

A brochure featuring basic information on recycling and solid waste management. It succinctly explains the problem of solid waste and identifies the solutions, encouraging both source reduction and recycling.

———. **Recycling in Federal Agencies.** Washington, DC: EPA, 1990. Free. EPA/530-SW-90-082.

———. **Recycling Works! State and Local Success Stories.** Washington, DC: EPA, 1989. 52p. Free. EPA/530-SW-89-014.

Examines 14 city and state recycling programs across America, each recycling different products. The type of program, a community overview, background facts, a program description, unique characteristics, and obstacles overcome are detailed for each locale.

———. **School Recycling Programs—A Handbook for Educators.** Washington, DC: EPA, 1990. 24p. Free. EPA/530-SW-90-023.

A brochure describing several options for setting up a school recycling program. It spotlights a number of successful efforts in this area and encourages the implementation of such programs to teach the concepts of source reduction and recycling.

———. **The Solid Waste Dilemma: An Agenda for Action.** Washington, DC: EPA, 1989. 70p. Free. EPA/530-SW-89-019.

The final report of the EPA's Municipal Solid Waste Task Force, which was to "fashion a strategy for improving the nation's management of municipal solid waste." This agenda offers specific suggestions for actions industry, citizens, and all levels of government can take to deal with solid waste problems. The systems approach of integrated solid waste management, rather than a focus on any one or two partial solutions, is emphasized.

———. **Yard Waste Composting: A Study of Eight Programs.** Washington, DC: EPA, 1989. 49p. Free. EPA/530-SW-89-038.

An informative introduction to yard waste composting, with case studies of eight communities from coast to coast. Additional contacts, references, and resources are listed as well.

Wolf, Nancy, and Ellen Feldman. **Plastics: America's Packaging Dilemma.** Covelo, CA: Island Press, 1991. 128p. $19.95. ISBN 1-55963-063-9.

An extensive treatment of the subject of plastics, including information on packaging, building materials, electrical products, adhesives, and related legislative and regulatory issues. This book also deals with controversies over the recyclability, degradability, and incineration of plastics.

Children's Books

The works listed in this section are specifically geared toward juvenile readers. Each can be read alone by a child or with a parent, teacher, or scout leader as a group project.

Appelhof, Mary, Mary Frances Fenton, and Barbara Loss Harris. **Worms Eat Our Garbage: Classroom Activities for a Better Environment.** Kalamazoo, MI: Flower Press, 1993. 232p. $21.95. ISBN 0-942256-05-0.

This author, well known for her expertise in the field of vermicomposting, takes this subject into the classroom. The result is a book full of lessons on recycling and composting, by way of worms, that entertains as it educates.

Chevat, Richard, and Joe Ewers. **Ready, Set, Recycle!** New York: Golden Press, 1993. 12p. $1.95. ISBN 0-307-10554-7.

One of the Golden Little Super Shape Book series.

The EarthWorks Group, eds. **50 Simple Things Kids Can Do to Save the Earth.** Kansas City, MO: Andrews and McMeel, 1990. 156p. $6.95. ISBN 0-8362-2301-2.

For 8- through 12-year-old children, but most of the ideas in this book can also be used by adults for family or class projects. Beginning with an explanation of environmental problems, this book offers information, experiments, and projects that teach children that they, too, can make a difference to the environment.

Elkington, John, Julia Hailes, Douglas Hill, and Joel Makower. **Going Green: A Kid's Handbook to Saving the Planet.** New York: Puffin Books, 1990. 111p. $8.95. ISBN 0-14-034597-3.

An excellent guide to "going green," from the editors of *The Green Consumer Letter* and the authors of *The Green Consumer.* Simple explanations and projects are included, along with a guide to the 3Rs, many pertinent facts, and a children's version of how to conduct a "green audit" at home, in school, and in the community. Color illustrations by British illustrator Tony Ross add to the well-organized and entertaining text.

Gibbons, Gail. **Recycle: A Handbook for Kids.** New York: Little, Brown, 1996. 32p. $4.95. ISBN 0-316-30943-5.

This brightly illustrated book tells all about what happens to all the things that are collected for recycling. Focusing on paper, glass, aluminum cans, and plastic, it explains what happens to each product and how it is recycled into new products.

Goodman, Billy. **A Kid's Guide to How to Save the Planet.** New York: Avon Books, 1990. 137p. $2.95. ISBN 0-380-76041-X.

This is another good book for children that provides an introduction to environmental problems and solutions. Filled with black-and-white photographs and illustrations, each chapter is divided into facts and simple explanations, followed by a "What You Can Do" section. One chapter is devoted to solid waste issues. Reading level 7.8.

McQueen, Kelly, and David Fassler. **Let's Talk Trash: The Kid's Book about Recycling.** New York: Waterfront Books, 1991. 169p. $14.95. ISBN 0-914525-19-0.

A good introduction to the subject of trash and recycling for younger children (kindergarten through third grade). This book incorporates the thoughts, ideas, and drawings of children in this age group in a broad discussion of trash, with an emphasis on recycling. The Environmental Law Foundation supported the writing of this paperback.

Savage, Candace, and Steve Beinicke. **Trash Attack: Garbage, and What We Can Do about It.** Buffalo, NY: Firefly Books, 1990. 56p. $9.95. ISBN 0-920668-73-9.

Part of the Earth Care Books series for children, geared toward readers in the fourth through sixth grades. The contents of this simple but thorough treatment of solid waste are divided into two sections—"The Problem" and "The Solutions"—with excellent color illustrations by Steve Beinicke.

Seltzer, Meyer. **Here Comes the Recycling Truck!** Morton Grove, IL: Albert Whitman, 1992. 32p. $14.95. ISBN 0-8075-3235-5.

This storybook follows the driver of a recycling truck as she collects glass, papers, and metals and takes them to the recycling center for preparation for recycling.

Journals

BioCycle
JG Press, Inc.
P.O. Box 351
Emmaus, PA 18049
(610) 967-4135
Monthly. $58.

Also known as the *Journal of Waste Recycling,* this magazine reports on new solutions to sludge and trash problems, focusing on composting, land application, materials recycling, and anaerobic digestion. Monthly information is provided on research findings, new equipment, financing options, health issues, government policies, innovative technologies, and marketing of treated residuals.

E Magazine
28 Knight Street
Norwalk, CT 06851
(203) 854-5559
(203) 866-0602
Subscription information: (815) 734-1242
Bimonthly. $20.

E/The Environmental Magazine is published by the nonprofit group Earth Action Network, Inc. It was founded in 1989 with the purpose of "acting as a clearinghouse of information, news, and commentary on environmental issues for the benefit of the general public." The magazine tries to both inform people and inspire activities to improve the Earth's environment.

P3, The Earth-Based Magazine for Kids
P3 Foundation
P.O. Box 52
Montgomery, VT 05470
5 issues per year. $18 (two-year subscription, or 10 issues).

P3 is an environmental magazine for seven- to ten-year-old children. Filled with colored pictures, short stories, and environmental news, it is a wealth of information for the younger set. The magazine also offers ideas for projects and activities that kids can accomplish without parental help. Subscription discounts are available to elementary schools and their libraries, nonprofit organizations, public libraries, and universities.

Recycling Today
GIE Inc., Publishers
4012 Bridge Avenue
Cleveland, OH 44113-3320
Toll-free subscription information number: 1-800-456-0707
Monthly. $30.

This "Business Magazine for Recycling Professionals," focuses on topics of interest to government and private recycling professionals. Regular departments include: "Business Watch," "News Watch," "Metal Watch," "Nonmetallics," and "C&D Debris."

Resource Recycling: North America's Recycling & Composting Journal
Resource Recycling, Inc.
P.O. Box 10540
Portland, OR 97210-0540
(503) 227-1319
Monthly. $47.

This periodical is a trade journal for those in the business of recycling and composting. It highlights various solid waste management programs around the country and delves into specific issues each month (e.g., citywide composting, office paper recycling, or aluminum foil recycling programs). Regular segments include: "Association Watch," "State/Province Watch," "Composting Update," "Market Update," "Programs in Action," "Equipment & Product News," and "Information Sources." Although geared to the industry, *Resource Recycling* also contains a wealth of information useful to the budding recyclist interested in forming or running a recycling program outside the household. Resource Recycling, Inc., also publishes two monthly newsletters: *Plastics Recycling Update* ($49 per year) and *Bottle/Can Recycling Update* ($55 per year).

Waste Age
4301 Connecticut Avenue, NW, Suite 300
Washington, DC 20008
(202) 244-4700
Toll-free subscription information number: 1-800-829-5411
Monthly. $55

This magazine is the monthly journal of the Environmental Industry Associations, and covers all aspects of the solid waste management industry. Regular feature sections include, "Collections," "Waste Transportation," "Municipalities," "Landfills," and "Recycling."

World Wastes
Intertec Publishing Corporation
6151 Powers Ferry Road
Atlanta, GA 30339-2941
(770) 955-2500
Monthly. $52.

This periodical, first introduced in 1958, is written for the waste management professional. It includes monthly information on the ever-evolving world of recycling, and also addresses the variety of different technologies available to handle every type of waste produced in the world. Regular departments include: "Finance," "Collection," "Technology," "Landfills," "Legislation," and "International." Editorial views are provided on recycling and composting, among other issues.

Solid Waste/Recycling Bibliographies

National Soft Drink Association. **Solid Waste & Recycling Bibliography.** Washington, DC: National Soft Drink Association, 1990. 31p. Free.

A pamphlet listing numerous state publications, waste exchanges, EPA documents, and scientific studies and research reports on specific commodities.
U.S. Environmental Protection Agency. **Bibliography of Municipal Solid Waste Management Alternatives.** Washington, DC: EPA, 1989. 10p. Free. EPA/530-SW-89-055.

A listing of approximately 200 publications available from government, industry, and environmental groups.

Newsletters

Information on most of the state recycling newsletters may be found in Chapter 6 under individual organization listings. The resources given here are national in scope.

The Green Business Letter
Tilden Press, Inc.
1519 Connecticut Avenue, NW
Washington, DC 20036
Toll-free subscription information number: 1-800-955-GREEN
Monthly. $127 (printed). $95 (e-mail).

This independently produced newsletter is aimed at helping companies make sound environmental ("green") choices that save money in the process. Discussion of recycling programs, products, and technologies is often included.

Recycled Products Business Letter
Environmental Newsletters, Inc.
11906 Paradise Lane
Herndon, VA 22071-1519
(703) 758-8436
Monthly. $149.

This newsletter is specifically geared toward helping companies that are trying to market recycled products.

Reusable News
Office of Program Management and Support (OPMS) OS-305
U.S. Environmental Protection Agency
401 M Street, SW
Washington, DC 20460
Quarterly. Free.

This newsletter, produced by the EPA's Office of Solid Waste, reports on the efforts of the EPA and other public and private groups at federal, state, or local levels to address the solid waste dilemma. The publication provides useful information about key issues and concerns in municipal solid waste management, including developments in recycling, source reduction, composting, landfilling, and procurement of recycled products. Additional information resources are listed in each issue.

Waste Age's Recycling Times
4301 Connecticut Avenue, NW, Suite 300
Washington, DC 20008
Toll-free subscription information number: 1-800-424-2869
Biweekly. $99.

This newspaper provides up-to-the-minute coverage of recycling markets and keeps subscribers informed on trends, legislation, and regulations affecting the recycling industry. It features in-depth reporting for the recycling professional.

Waste Reduction Tips
Environmental Newsletters, Inc.
11906 Paradise Lane
Herndon, VA 22071-1519
(703) 758-8436
Monthly. $97.

This newsletter is written to individuals and companies, informing its audience on ways to reduce waste generation at its source in the home or office.

Curriculum Guides and Educational Materials

Most curriculum guides on solid waste and recycling can be obtained through state agencies and regional offices of the EPA, which are listed in Chapter 6. The organizations listed here have environmental curriculum guides that are specific to their individual areas of interest.

American Plastics Council (APC)
1275 K Street, NW
Washington, DC 20005
1-800-2-HELP-90

APC has two curricula kits entitled *Plastic in Our World,* one targeted at grades K–6 and the other for grades 7–12. The kits include activities booklets, a How to Set Up a School Recycling Program guide, and additional brochures for the science teacher or curriculum-development specialist. Single copies are free to teachers. The *Hands on Plastics Kit: A Scientific Investigation Kit* is designed by APC to help middle school–level science teachers and their classes explore the world of plastics through a variety of lessons about chemical structures, resin identification codes, the many different forms of plastics, and the science of recycling. The kit includes samples of plastics, learning cycle activities and extensions, and background information. $10.

Association of Vermont Recyclers (AVR)
P.O. Box 1244
Montpelier, VT 05601-1244
(802) 229-1833

The *AVR Teacher's Resource Guide for Solid Waste and Recycling* is a nationally recognized interdisciplinary program. Intended for grades K–12, the 340-page resource guide is full of inquiry-based activities. (Costs vary: $20 for in-state members; $25 for in-state nonmembers; $40 for out-of-state members; $45 for out-of-state nonmembers.) AVR also produces a resource guide for how to establish a school lunch composting program. *Scraps to Soil—A "How To" Guide for School Composting* includes both student and teacher pages, extension activities, resource materials, and contacts. (Free to teachers in-state; $10 out-of-state.)

Glass Packaging Institute (GPI)
1627 K Street, NW, Suite 800
Washington, DC 20006
(202) 887-4850

The Great Glass Caper is a teacher's guide on recycling geared toward elementary-level students that has been produced and distributed for the past ten years or so. Additionally, a new teacher's guide is now available from GPI featuring a tiered range of activities for grades 1–12. The kit includes a colorful "interactive" poster for use in the classroom. Free.

Institute of Scrap Recycling Industries, Inc. (ISRI)
1325 G Street, NW, Suite 1000
Washington, DC 20005-3104
(202) 737-1770
(202) 626-0900 (FAX)

ISRI produces a school program called *The Scrap Map* for grades K–6. The materials offered include a brochure for students, a Teacher's Kit, and a classroom package with brochures, posters, a video, and additional literature for presentation. For nonprofits or schools, the brochures cost $15 for a package of 30, the Teacher's Kit is $5, and the classroom presentation package is $80.

Keep America Beautiful, Inc.
Attn: KAB Materials
Mill River Plaza
9 West Broad Street
Stamford, CT 06902
(203) 323-8987

Waste in Place, a sequential curriculum supplement for teachers of grades K–6, deals with proper waste handling. It is interdisciplinary in approach, with activities relating to math, language arts, science, civics, and social studies. $40. *Waste: A Hidden Resource* is written for teachers of students in grades 7–12 and is designed to help teenagers develop responsible behaviors and attitudes toward the environment. The interdisciplinary course deals with the various waste disposal solutions. $50.

National Solid Wastes Management Association (NSWMA)
Publications Department
Attn: Scott Jones
1730 Rhode Island Avenue, NW, Suite 1000
Washington, DC 20036
(202) 659-4613
Toll-free order line: 1-800-424-2869

NSWMA produces a number of publications geared specifically toward educating children about solid waste:

- *Garbage Then and Now,* a timeline approach to the history of solid waste in the world, is a colorful and informative brochure. $1.50 (may be ordered in packages of 25 for $18.75, 50 for $29.50, and 100 for $45). #003118.
- In the *Walt Wastenot Activity Book,* a waste-watching squirrel named Walt teaches children about recyling, hazardous wastes, and the collection and transfer of solid waste. Puzzles, mazes, and other fun games to enhance learning are included. $4.50. #001039.
- The *Mini-Page on Trash,* a Sunday newspaper comics supplement, was first published across the country on the 1989–1990 New Year's weekend. Using fun facts and illustrations, it examines issues such as what we throw away, what and why we recycle, and how landfills work. $1. #90001.

RCRA Information Center
OS-305
U.S. Environmental Protection Agency
401 M Street, SW
Washington, DC 20460
Hotline number: 1-800-424-9346
E-mail: rcra-docket@epamail.epa.gov

The center produces *Recycle Today!,* a curriculum about recycling and other solid waste issues for grades kindergarten through 12. This program, developed by the EPA's Office of Solid Waste with advice from several national educational organizations, teaches language, science, math, art, and social studies skills as it educates about recycling. It consists of four publications, each of which may be ordered individually by its reference number: *Recycle Today! Curriculum* (EPA/530-SW-90-005); *How-To Handbook* (EPA/530-SW-90-023); *Comic Book* (EPA/530-SW-90-024); *Poster* (EPA/530-SW-90-010).

Articles

Ackerman, Frank. "Solid Waste: The Hidden Utility." *BioCycle* (August 1990): 84–85.

Alexander, Michael. "Old Newspapers and Magazines: Is There More Room for Recovery?" *Resource Recycling* (April 1996): 40–48.

American Plastics Council. "What Industry Is Doing." Washington, D.C.: American Plastics Council, June 1996.

Andress, Carol. "Innovative Funding for Recycling Programs." *BioCycle* (August 1990): 50–53.

Apotheker, Steve. "Changing the Look of Recovered Paper Processing." *Resource Recycling* (April 1996): 17–27.

———. "Glass Containers: How Recyclable Will They Be in the 1990s?" *Resource Recycling* (June 1991): 25–32.

———. "Office Paper Recycling: Collection Trends." *Resource Recycling* (November 1991): 44–52.

Bloyd-Peshkin, Sharon. "Taming the Waste Stream in Your Office." *Vegetarian Times* (July 1990): 54–61.

Breen, Bill. "Selling It! The Making of Markets for Recyclables." *GARBAGE* (November/December 1990): 44–49.

Brewer, Gretchen. "Plastic Bottles Close the Loop." *Resource Recycling* (May 1991): 88–95.

Broughton, Anne Clair. "Beyond Precious Metal Refining." *Recycling Today* (February 1996): 40–50.

———. "Buried Alive: The Garbage Glut." *Newsweek* (27 November 1989): 66–76.

Christrup, Judy. "Duped by Plastics." *Greenpeace* (September/October 1989): 18.

Daniel, Joseph E. "The Glossy Truth of Recycled Paper." *Buzzworm* (July/August 1990): 20–21.

Davis, Alan, and Susan Kinsella. "Recycled Paper: Exploding the Myths." *GARBAGE* (May/June 1990): 48–54.

DeYoung, R. "Recycling as Appropriate Behavior: A Review of Survey Data from Selected Recycling Education Programs in Michigan." *Resources, Conservation, and Recycling* (1990): 253–266.

Easterbrook, Gregg. "Cleaning Up Our Mess." *Newsweek* (July 24, 1989): 26–42.

Egan, Katherine. "Westchester County, N.Y., Is Testing New Glass Sorting Equipment." *Recycling Times* (January 20, 1997): 7.

Emm, Joseph. "Hazardous Household Waste." *Home* (October 1996): 140–144.

Environmental Action, eds. "Recycling Plastics: A Forum." *Environmental Action* (July/August 1988): 21–25.

Erkenswick, Jane L. "Office Paper Recycling: A Look at the Ledger Grades." *Resource Recycling* (November 1991): 64–68.

Farmanfarmaian, Roxane. "Saving the Environment." *McCall's* (January 1991): 65–69, 139.

Foote, Knowlton C. "Setting Up Apartment Recycling: Some Things to Remember." *Resource Recycling* (April 1996): 49–57.

Frosch, Robert A., and Nicholas E. Gallopoulos. "Strategies for Manufacturing." *Scientific American* (September 1989): 144–152.

Gibson, Susan. "Manufacturers Eye Design for Recycling Options." *Recycling Today* (15 October 1990): 121–124.

Glenn, Jim. "After the Crisis." *BioCycle* (July 1990).

———. "The State of Garbage in America." *BioCycle* (March 1990): 48–53.

Graham, D. Douglas. "C&D Waste Opens a Window of Opportunity." *World Wastes* (April 1996): 28–34.

Grassy, John. "Bottle Bills: Headed for a Collision at Curbside?" *GARBAGE* (January/February 1992): 45–47.

Grogan, Peter L. "Beyond Recycled Fiber Content." *BioCycle* (June 1990): 85.

———. "Grassroots Grass Clippings." *BioCycle* (July 1990): 80.

Guttentag, Roger M. "Recycling in Cyberspace: The Searchers." *Resource Recycling* (April 1996): 76–77.

Hamilton, Shaan Kervis. "Optimizing the Collection of Glass Containers." *Resource Recycling* (July 1995).

Harler, Curt. "Keeping It All in the Family." *Recycling Today* (March 1991): 50–53.

Hasek, Glenn. "Hotels Keeping Watch on Waste." *Resource Recycling* (January 1991): 56–60.

Heumann, Jenny M. "Tempe, Ariz., Stops Glass Collection." *Recycling Times* (January 20, 1997): 4.

———. "Will a Bottle Bill Pass in Georgia?" *Waste Age's Recycling Times* (January 20, 1997): 14.

Hong, Peter, and Dori Jones Yang. "Cutting the Trash Heap Down to Size." *Business Week* (30 September 1991): 100L.

Keene-Osborn, Sherry. "Recycling: Will the Cycle Be Unbroken?" *Colorado Business* (April 1991): 34–42.

Kourik, Robert. "What's So Great about Seattle?" *GARBAGE* (November/December 1990): 24–31.

Levetan, Steve. "Can Government *R* Dictate Who Handles Recyclables?" *Recycling Today* (January 1991): 8.

Logsdon, Gene. "Agony and Ecstasy of Tire Recycling." *BioCycle* (July 1990): 44–85.

Long, Lynda. "Smart Budgeting to Promote Recycling." *BioCycle* (March 1990): 54–55.

Luoma, Jon R. "Trash Can Realities." *Audubon* (March 1990): 86–97.

Marshall, Jonathon, and Daniel Vasquez. "Recycling Goals May Need Recycling." *San Francisco Chronicle* (17 December 1991): 1.

Mayville, Gail. "Paper Recycling at the Office." *E Magazine* (January/February 1991): 48.

Moll, Lucy. "The Plastics Rap." *Vegetarian Times* (February 1989): 40–51.

Monk, Randall. "Municipal Composting Comes of Age." *World Wastes* (October 1991): 34–40.

Montanari, Richard. "Getting Connected." *Recycling Today* (April 1991): 74–77.

Newton, James W., and W. J. Rohwedder. "Environmental Computer Networking." *E Magazine* (March/April 1990): 45–47.

Pardue, Leslie. "Plastics, Plastics Everywhere." *E Magazine* (November/December 1991): 48–50.

Patterson, Gene E. "How Much Does Recycling Reduce the Waste Stream?" *BioCycle* (July 1990): 46–49.

Porter, J. Winston. "Let's Go Easy on Recycling Plastics." *Philadelphia Inquirer* (6 August 1991): 2.

Powell, Jerry. "All Types and Sizes Available: Recent Scrap Tire Recycling Legislation." *Resource Recycling* (December 1990): 60–64.

Powell, Jerry, and McEntee, Ken. "Office Paper Recycling: Existing and Emerging Markets." *Resource Recycling* (November 1991): 53–57.

Rathje, William L. "The History of Garbage." *GARBAGE* (September/October 1990): 32–39.

———. "Recycled Phone Books Find Home in Brewery Sludge." *BioCycle* (November 1991): 62.

———. "Rubbish." *The Atlantic Monthly* (December 1989): 99–109.

Rathje, William, and Cullen Murphy. "Five Major Myths about Garbage, and Why They're Wrong." *Smithsonian* (July 1992).

Raymond, Michele. "Associations Don't Expect Much Action in 1997." *State Recycling Laws Update* (December 1996). World Wide Web: http://www.raymond.com/recycle.

———. "Cities Are Livid over PET Problems." *State Recycling Laws Update* (December 1996). World Wide Web: http://www.raymond.com/recycle.

Redd, Adrienne. "Pennsylvania Mirrors Recycling across America." *World Wastes* (April 1996): 44–47.

Rembert, Tracey C. "Bananas Save Trees." *E Magazine* (January/February 1997): 23.

———. "Dam Those Tires!" *E Magazine* (January/February1997): 21–22.

Reutlinger, Nancy, and Dan de Grassi. "Household Battery Recycling: Numerous Obstacles, Few Solutions." *Resource Recycling* (April 1991): 24–29.

Rockland, Dr. David B. "Paper Recycling: Fact from Fiction." *Home Mechanix* (October 1991): 15–16.

Ruffer, Deanna L. "Life after Flow Control: Assuring the Economic Viability of Local Government Solid Waste Management Systems." *Waste Age* (January 1997): 73–78.

Schut, Jan H. "McDonald's Move Jolts PS Recycling but Won't Halt Numerous Ventures." *Plastics Technology* (January 1991): 99–101.

Striano, Elizabeth. "NYC Plans for Closure of Sole Landfill." *Waste Age* (January 1997): 24–29.

Tedeschi, Bruno. "Recycling at GARBAGE." *GARBAGE* (March/April 1990): 50–51.

Thayer, Ann M. "Solid Waste Concerns Spur Plastic Recycling Efforts." *Chemical & Engineering News* (January 30, 1989): 7–15.

Trank, Andrea. "Green Paper." *In Business* (November/December 1990): 36–38.

University of Illinois at Chicago. "Composting." *Solid Waste Management Newsletter*, vol. 10 (October 1993).

Watson, Tom. "Heading for the Hills: Rural Curbside Recycling." *Resource Recycling* (November 1991): 23–26.

Westerman, Martin. "Restaurants Recycle." *Resource Recycling* (January 1991): 78–83.

Wirka, Jeanne. "A Plastics Packaging Primer." *Environmental Action* (July/August 1988): 17–20.

Young, John E. "Tossing the Throwaway Habit." *World Watch* (May/June 1991): 26–33.

Videocassettes

Most audiovisual resources in the fields of solid waste and recycling are produced by government agencies and trade organizations. It is best to remember that the videos available through trade organizations are generally aimed at promoting their visions of the industry and encouraging the use of their products. However, there are some excellent educational videos for juvenile audiences (although adults can often benefit from this type of presentation as well), and those distributed by government agencies usually include sound background information on solid waste issues.

In many cases, state government agencies and regional offices of the Environmental Protection Agency produce videos specific to the areas they serve. It is best to consult such local offices (see Chapter 6) to see what is available because these resources may not be listed here. Also, organizations that publish solid waste management curricula often produce videotapes as well (as mentioned earlier in this chapter), and many state agencies produce videos that are available for rental or sale.

Solid Waste/Recycling/Composting

Changing Skylines: The Garbage Crisis
Type: VHS videotape
Length: 29 minutes
Date: 1990
Cost: Free for public education
Source: EPA Region 8
8HWM-RI
999 18th Street
Denver, CO 80202-2405

Complete Home Composting Video Guide
Type: VHS videotape
Date: 1996
Cost: $14.99
Source: Tapeworm Video
12229 Montague Street
Arleta, CA 91331
(818) 896-8899

Composting for the 90's
Type: VHS videotape
Length: 50 minutes
Date: 1995
Cost: $24.99
Source: Eternity Road Productions
P.O. Box 1943
Jackson, MI 49204
(517) 787-6043

Earth Aid: Recycling
Type: VHS videotape
Length: 17 minutes
Date: 1995
Cost: $14.99
Source: Viejo Publications
6008 SW 21st
Portland, OR 97201
(503) 246-3790

Garbage: The Movie—An Environmental Crisis
Type: VHS videotape
Length: 25 minutes
Date: 1990
Source: Churchill Films
12210 Nebraska Avenue
Los Angeles, CA 90025

This award-winning film takes a neatly organized look at solid waste problems in the United States. Using the "Garbage Barge" of Islip, New York, as a central theme, the producers explain the modern garbage dilemma and offer steps toward solutions. The 3Rs are demonstrated in this positive and informative film, and the viewer learns how recycling saves energy, how high school students can start their own recycling programs, and how individuals and communities are composting yard wastes.

In Partnership with Earth: Pollution Prevention for the 1990s
Type: VHS videotape
Length: 58 minutes
Date: 1990
Cost: $40
Source: Versar Inc.
6850 Versar Center
P.O. Box 1549
Springfield, VA 22151

John Denver hosts and narrates this film, which reviews environmental achievements over the past 20 years. Nature's ways of preventing pollution through recycling are compared with those developed by men and women. With a positive look at the progress in recycling made by large corporations such as IBM and Eastman Kodak, this video is aimed at general audiences, ages 12 and older.

Mister Rogers' Recycling Video
Type: VHS
Length: 30 minutes
Date: 1992
Cost: $19.95
Source: Keep America Beautiful, Inc.
1010 Washington Boulevard
Stamford, CT 06901
(203) 323-8987

Produced by PBS, this video is ideal for libraries and preschools.

Other videotapes available from Keep America Beautiful:

Beyond the Bins—Sustaining a Recycling Program in Your Community
Length: 15 minutes
Date: 1993
Cost: $7.50

Describes the steps that take place beyond the collection of household recyclables.

Recycling Realities: A National Town Meeting
Length: 120 minutes
Date: 1993
Cost: $20

This is a live taping of Keep America Beautiful's interactive videoconference in 1993 that addressed the perceptions versus the realities of recycling, as well as other options for community waste management (emphasizing the integrated approach).

Recycle It Yourself
Type: VHS videotape
Length: 37 minutes
Date: 1990
Cost: $29.95
Source: Aylmer Press
P.O. Box 2735
Madison, WI 53701

This video is more of an arts and crafts presentation than an environmental statement. Although it demonstrates how different items (e.g., coasters, necklaces, and paper) can be made at home with recycled materials, it offers no comments on an individual's need to recycle or to be concerned about the environment.

Recycling Paper and Cardboard
Type: VHS videotape
Length: N/A
Date: 1990
Cost: $28
Source: Troll Associates (Library)
Junk Ecology Series
Troll Communication
100 Corporate Drive
Mahwah, NJ 07430
(201) 529-4000
Toll-free: 1-800-929-8765

Recycling Plastic Throwaways
Type: VHS videotape
Length: NA
Date: 1990
Cost: $25.20
Source: Troll Associates (Library)
Junk Ecology Series
Troll Communication
100 Corporate Drive
Mahwah, NJ 07430
(201) 529-4000
Toll-free: 1-800-929-8765

Reducing, Reusing, and Recycling: Environmental Concerns
Type: VHS videotape
Length: 20 minutes
Date: 1990
Cost: $89
Source: Rainbow Educational Video
170 Keyland Court
Bohemia, NY 11716

Recommended for six- to eight-year-old children, this film defines basic terms such as "solid waste," "decompose," and "nonrenewable resource" and intelligently discusses the 3Rs. An accompanying teacher's guide and worksheet can further the learning process.

The Rotten Truth
Type: VHS videotape
Length: 30 minutes
Date: 1990
Cost: $19.95 (including a 32-page teacher's guide)
Source: Sunburst Communications
101 Castleton Street
P.O. Box 100
Pleasantville, NY 10570-9963
Toll-free number for orders: 1-800-321-7511

This is an excellent tour through the world of trash with child actress Stephanie Yu from the PBS series *3–2–1 CONTACT*. The tape emphasizes the fact that "you can't make nothing from something" when it comes to trash. Landfills, incineration, composting, and recycling are covered, as well as source reduction (with a lively song on excessive packaging by "The Wrapper").

Save the Earth: A How-To Video
Type: VHS videotape
Length: 60 minutes
Date: 1990
Cost: $19.95
Source: Tri-Coast International
1020 Pico Boulevard
Santa Monica, CA 90405

Although it ignores economic policy, global politics, and sustainable development issues, this video does a good job of advocating the 3Rs. Few environmental activist tactics, such as sending excess packaging back to the manufacturer, are suggested. The video contains the warning, "Life as we know it cannot continue."

The Earth Day Special
Type: VHS videotape
Length: 99 minutes
Date: 1990
Cost: $9.95
Source: Warner Home Video
4000 Warner Boulevard
Burbank, CA 91522

A made-for-television special with major motion picture stars such as Bette Midler, Robin Williams, and Meryl Streep, this is an entertaining but lengthy look at conservation of resources and energy. Several community recycling projects are shown.

Tinka's Planet
Type: VHS videotape
Length: 12 minutes
Date: 1990
Cost: $24.94, home use; $55, public performance
Source: The Video Project
5332 College Avenue, Suite 101
Oakland, CA 94618

Geared toward children in the primary grades, this video features a child who talks to a garbage collector to find out how to help the planet. After learning about landfills, Tinka learns how to recycle cans, glass, plastic, and paper. A catchy rap music song summarizes the main tips.

Why Waste a Second Chance?
Type: VHS videotape
Length: 12 minutes
Date: 1990
Cost: $80 nonmembers; $50 members
(two-week rental: $40 nonmembers, $20 members)
Source: National Association of Towns and Townships
1522 K Street, NW, Suite 730
Washington, DC 20005

This motivational video focuses on common solid waste disposal problems and highlights several recycling alternatives being used with success by small communities throughout the country. This video is designed to help community leaders and other concerned citizens generate interest and support for recycling. It is intended to be used with the accompanying guidebook (see the entry on National Association of Towns and Townships in Chapter 6).

Plastics

The Busy, Busy Planet
Type: VHS Video
Length: 11 minutes
Date: 1993
Cost: Single copy free to educators; $10 per copy for others
Source: American Plastics Council
1275 K Street, NW
Washington, DC 20005
1-800-2-HELP-90

This video proposes the many resource conservation benefits that plastics provide in everyday life.

Convenience Recycled
Type: VHS videotape
Length: 13 minutes
Date: 1991
Cost: Free
Source: Polystyrene Packaging Council, Inc.
1025 Connecticut Avenue, NW, Suite 515
Washington, DC 20036

Examining the industry view of the paper versus plastic controversy, this video appears to be a response to efforts to limit or ban polystyrene products. It promotes the recycling of plastics, especially polystyrene, and shows what products can be manufactured from this process. Profiles of several local polystyrene recycling programs around the country are also included.

Everybody's Talking about the Environment
Type: VHS videotape
Length: 13 minutes
Date: 1990
Cost: Free
Source: Foodservice & Packaging Institute, Inc.
1901 N. Moore Street, Suite 1111
Arlington, VA 22209

This film presents the National Polystyrene Recycling Company's views on recycling their products, as opposed to disposing of them. It also shows the variety of products that can be made from recycled polystyrene.

Foodservice Disposables: Should I Feel Guilty?
Type: VHS videotape
Length: 12 minutes
Date: 1991
Cost: Free
Source: Foodservice & Packaging Institute, Inc.
1901 N. Moore Street, Suite 1111
Arlington, VA 22209

From an industry viewpoint on the controversy over reusables versus disposables, this film discusses issues such as environmental trade-offs, natural resource conservation, and litter. The focus is on the fast-food habits of Americans.

Polystyrene: Solid Waste Solutions
Type: VHS videotape
Length: 14 minutes
Date: 1990
Cost: Free
Source: Polystyrene Packaging Council, Inc.
1025 Connecticut Avenue, NW, Suite 508
Washington, DC 20036

This video provides more industry information on polystyrene recycling.

Miscellaneous Other Resources

Environmental Protection Agency. ***Municipal Solid Waste Factbook, ver. 3.0.*** Computer software. EPA, 1996. Requires Windows 3.1 or later. EPA530-C-96–001.

This software, in Windows™ format, is a veritable treasure chest of information about solid waste management in the United States. There are also a few screens that deal with international waste issues. The program contains statistics, graphs, charts, and tables about all facets of the waste industry, from recycling to combustion. This electronic disk reference manual is free upon request through the RCRA hotline: 1-800-424-9346.

8 Online Resources

In 1992, when the first edition of this work was published, computer networks were a newly emerging technology as far as the "normal" person was concerned. The field has changed considerably since that time, and the Internet has all but become a household commodity. As a result, this chapter has been added to facilitate Internet searches in the areas of recycling and solid waste subject matter. Both e-mail addresses and World Wide Web pages are listed wherever possible.

Additionally, at the 1996 annual conference of the National Recycling Coalition, the State Recycling Organization Council began an effort to get all state recycling organizations on the Internet by the end of 1997. Thus, by the time this book is in print, more Internet state recycling resources should be available.

Internet search engines, such as Alta Vista, YAHOO!, WebCrawler, and Lycos, will also yield hundreds of site matches from the word "recycling," ranging from companies that will recycle assorted products to animated cartoons addressing recycling education. These can be accessed directly by using keywords (e.g., recycling, solid waste management, or packaging). The resources listed here are limited to government agencies (Canadian and U.S.), national organizations, product-specific organizations, and periodicals. One Internet bulletin board area, *Recycle Talk* (part of the Global Recycling Network), is also included because of its excellent ability to provide up-to-the-minute, superbly researched information on almost every aspect of recycling. The Global Recycling Network (GRN) is a network solely devoted to the business of the recycling industry.

Federal Agencies

"Pay-as-You-Throw" Information
E-mail: ferland.henry@epamail.epa.gov

RCRA Information Center
E-mail: rcra-docket@epamail.epa.gov

U.S. Environmental Protection Agency
WWW: http://www.epa.gov

State Agencies

Alabama Department of Environmental Management
WWW: http://alaweb.asc.edu/govern.html

Alaska Department of Environmental Conservation
WWW: http://www.state.ak.us/

Arkansas Recycling Coalition
E-mail: recyclearc@aristotle.net

California Department of Conservation
WWW: http://www.consrv.ca.gov/

California Integrated Waste Management Board
WWW: http://www.ciwmb.ca.gov

California Resource Recovery Association
E-mail: CRRA@aol.com

Northern California Recycling Association
E-mail: akrakowski@igc.apc.org

Colorado Department of Public Health and the Environment
WWW: http://www.state.co.us/gov_dir/

Colorado Governor's Office of Energy Conservation
WWW: http://www.state.co.us/gov_dir/energy_gov.html
E-mail: oec@csn.net

Delaware Department of Natural Resources and Environmental Control
WWW: http://gis01.dnrec.state.de.us/

Florida Department of Environmental Protection
WWW: http://www.dep.state.fl.us/index.html

Florida Division of Waste Management
WWW: http://www.dep.state.fl.us./waste/index.htm

RecycleFlorida Today, Inc.
WWW: http://www.enviroworld.com/Resources/RFT.html
E-mail: recyclefl@aol.com

Georgia Department of Natural Resources
WWW: http://www.state.ga.us/

Hawaii Department of Health
WWW: http://www.hinc.hinc.hawaii.gov

Idaho North Central District Health Department
WWW: http://www.idwr.state.id.us/oneplan/waste/handle.htm

Illinois Environmental Protection Agency
WWW: http://www.epa.state.il.us/

Illinois Recycling Association
E-mail: ILRecycle@aol.com

Indiana Recycling Coalition
WWW: http://www.papertrail.com/irc/
E-mail: recycling@in.net

Iowa Department of Natural Resources
WWW: http://www.igsb.uiowa.edu/dnr_www.htm

Iowa Recycling Association
E-mail: munderwo@ided.state.ia.us

Recycle Iowa
WWW: http://www.RecycleIowa.org

Louisiana Department of Environmental Quality
WWW: http://www.deq.state.la.us/oshw/oshw.htm

Louisiana DEQ Recycling
E-mail: john_r@deq.state.la.us

Maine Department of Environmental Protection
WWW: http://www.state.me.us/dep/mdephome.html

Maine State Planning Office
WWW: http://www.state.me.us/spo

Maryland Department of Natural Resources
WWW: http://gacc.com:82/

Maryland Department of the Environment
WWW: http://www.mde.state.md.us

Maryland Recyclers Coalition
WWW: http://www.marylandrecyclers.org
E-mail: mrc@mdassn.com

Massachusetts Department of Environmental Protection
WWW: http://www.state.ma.us/dep/bwp/dswm/dswmhome.htm

MassRecycle (Massachusetts)
E-mail: massrecy@aol.com

Michigan Department of Environmental Quality
WWW: http://www.deq.state.mi.us/
E-mail: doroshkl@deq.state.mi.us

Michigan Department of Natural Resources
WWW: http://www.dnr.state.mi.us/

Minnesota Pollution Control Agency
WWW: http://www.pca.state.mn.us

New Hampshire Department of Environmental Services
WWW: http://www.state.nh.us/des/descover.html

Association of New Jersey Recyclers
E-mail: anjr@ifu.net

New York Association for Reduction, Reuse, and Recycling
WWW: http://www.recycle.net/recycle/assn/nysarrr

North Carolina Division of Waste Management
WWW: http://www.watenot.ehnr.state.nc.us/

North Carolina Recycling Association
WWW: http://www.recycle.net/recycle/ncra/index.html
E-mail: ncrecycles@aol.com

North Dakota Department of Health
WWW: http://165.234.109.13/ndhd/index.html

Association of Ohio Recyclers
E-mail: pjsmith@prgone.com

Ohio Department of Natural Resources
WWW: http://www.dnr.ohio.gov/odnr/recycling

Ohio Environmental Protection Agency
WWW: http://arcboy.epa.ohio.gov/oepa.html

Ohio Recycling Information and Communication System
WWW: http://han6.hannah.com/odnr/mainmenu.html

Oklahoma Department of Environmental Quality
WWW: http://www.deq.state.ok.us/home.html

Association of Oregon Recyclers
E-mail: aor@mindspring.com

Commonwealth of Pennsylvania Department of Environmental Protection
WWW: http://www.dep.state.pa.us/dep/

Pennsylvania Resources Council
WWW: http://www.voicenet.com/~rusw/prc/
E-mail: imperato@prc.org

Rhode Island Department of Environmental Management, OSCAR Program
E-mail: oscar@oscar.state.ri.us

South Dakota Department of the Environment and Natural Resources
WWW: http://www.state.sd.us/state/executive/denr/denr.html

Tennessee Solid Waste Education Project
WWW: http://triton.rtd.utk.edu/eercwww/home.html

Texas Natural Resources Commission
WWW: http://tnrcc.texas.gov/

Utah Department of Environmental Quality
WWW: http://www.eq.ex.state.ut.us/eqoas/p2/p2_home.htm

Association of Vermont Recyclers
WWW: http://www.sover.net/~recycle/index.html
E-mail: recycle@sover.net

Washington Refuse and Recycling Association
E-mail: wrralr@aol.com

Washington State Department of Ecology
WWW: http://www.wa.gov/ecology

Washington State Recycling Association
E-mail: WSRA@aol.com

Associated Recyclers of Wisconsin
E-mail: cow@mailbag.co

Wyoming Department of Environmental Quality
WWW: http://deq.state.wy.us/ms.htm
E-mail: dhogle@missc.state.wy.us

Canadian Recycling Resources

Association of Municipal Recycling Coordinators
E-mail: amrc@web.apc.org

British Columbia Environment
WWW: http://ssbwww.env.gov.bc.ca/

Canadian Association of Recycling Industries
E-mail: cari2@cyclor.ca

Ecology Action Centre of Nova Scotia
E-mail: ip-eac@ccn.cs.dal.ca

Edmonton Recycling Society
WWW: http://www.mennonitecc.ca/mcc/pr/1994/04–07/4.html

Environment Canada
WWW: http://www.doe.ca/envhome.html

Greater Vancouver Regional District (BC)
WWW: http://www.grvd.bc.ca/

Province of BC Environment
WWW: http://ssbuxt.env.gov.bc.ca/

Recycling Council of Alberta
E-mail: rca@cadvision.com

Recycling Council of Manitoba
WWW: http://www.freenet.mb.ca/iphome/r/recycle/services.html

Recycling Council of Ontario
WWW: http://www.web.apc.org/rco/
E-mail: rco@web.apc.org

National Associations and Organizations

Air & Waste Management Association
WWW: http://www.awma.org/

Center for Neighborhood Technology
WWW: http://www.cnt.org

Center for Resourceful Building Technology
E-mail: crbt@aol.com

Environmental Action Foundation
WWW: http://www.econet.apc.org/eaf
E-mail: eaf@igc.apc.org

Environmental Defense Fund
WWW: http://www.edf.org
E-mail: webmaster@edf.org

Environmental Industry Associations
WWW: http://www.envasns.org

Environmental Production Guide
WWW: http://www. Epg.org

Fibre Box Association
WWW: http://www.loa.com/corrugated/

Flexible Packaging Association
E-mail: fpa@ssn.com

Foodservice & Packaging Institute, Inc.
WWW: http://www.fpi.org/fpi

Global Recycling Network
WWW: http://www.grn.com/grn/org/org-t.html

INFORM, Inc.
E-mail: Inform@igc.apc.org

Institute for Local Self-Reliance
WWW: http://www.great-lakes.net:2200/0/partners/ILSR/ILSRhome.html
E-mail: ilsr@igc.apc.org

Materials for the Future Foundation
E-mail: mff@igc.apc.org

National Waste Prevention Coalition
E-mail: twatson@eskimo.com

Natural Resources Defense Council
WWW: http://www.nrdc.org/nrdc/

Renew America
WWW: http://www.crest.org/renew_america
E-mail: renewamerica@igc.apc.org

Solid Waste Association of North America
WWW: http://www.swana.org

Waste Prevention Information Exchange
E-mail: wpinfoex@mrt.ciwmb.ca.gov

Worldwatch Institute
WWW: http://www.worldwatch.org
E-mail: worldwatch@worldwatch.org

Products

Glass

Glass Packaging Institute
WWW: http://www.gpi.org

Metals

American Iron & Steel Institute
WWW: http://www.steel.org

Can Manufacturers Institute
WWW: http://www.cancentral.com

The Copper Page
WWW: http://www.copper.org

Reynolds Aluminum Recycling
WWW: http://www.rmc.com/divs/recycle.html

Steel Recycling Institute
WWW: http://www.recycle-steel.org
E-mail: sri@recycle-steel.org

Plastics

American Plastics Council
WWW: http://www.plasticsresource.com

Center for Plastics Recycling Research
E-mail: rsaba@gandalf.rutgers.edu

National Association for Plastic Container Recovery
WWW: http://www.napcor.com
E-mail: NAPCOR@aol.com

Plastic Bag Association
WWW: http://www.plasticbag.com/pba.html

Plastic Bag Information Clearinghouse
E-mail: pbainfo@aol.com

The Plastics Group of America
WWW: http://www.plasticsgroup.com

The Plastics Network
WWW: http://www.plasticsnet.com

Polystyrene Packaging Council
WWW: http://www.polystyrene.org

Recycled Plastics Marketing
E-mail: rpm@aa.net

Society of the Plastics Industry
WWW: http://www.socplas.org

Composting

City Farmer
WWW: http://www.cityfarmer.org/

Compost Resource Page
WWW: http://www.oldgrowth.org/compost

Composting Council
E-mail: comcouncil@aol.com

Cosmos Red Worms
http://www.alcasoft.com/cosmos

Ecology Action Centre
http://www.cfn.cs.dal.ca/Environment/EAC/eac-internet-resources.html

Rodale Institute
WWW: http://www.envirolink.org/seel/rodale/

RotWeb
WWW: http://www.indra.com/~topsoil/Compost_Menu.html

Worm World
WWW: http://www.nj.com/yucky/worm/

Miscellaneous

Allied Signal Carpet Recycling
WWW: http://grn.com/grn/home/alliedsignal.htm

Computer Recycling Center
WWW: http://www.perspective.com/crc/

Construction Materials Recycling Association
WWW: http://351.39.1.2/CMRA/INFO

Eastman Kodak Company
WWW: http://www.kodak.com

Global Recycling Network
WWW: http://grn.com

Goodwill Recycling
WWW: http://www.ocgoodwill.org/gii/recycle.html

GreenLites Lamp Recycling
WWW: http://www.greenlites.com

Institute of Scrap Recycling Industries
WWW: http://www.isri.org

International Tire & Rubber Association
WWW: http://www.itra.com
E-mail: itra@aol.com

The Internet Consumer Recycling Guide
WWW: http://www.best.com:80/~dillon/recycle/

National Oil Recyclers Association
WWW: http://www.webcom.com/-infoserve/NORA

National Soft Drink Association
WWW: http://www.NSDA.org
E-mail: mariec@nsda.com

Recycler's World
WWW: http://www.recycle.net/recycle/index.html

Rubber Waste Technology
WWW: http://www.interpublish.com/rwt/index.html

Tire Retread Information Bureau
E-mail: retreads@aol.com

Waste Watch Center
E-mail: wastewatch@aol.com

Wood Recycling
WWW: http://grn.com/grn/home/wri.htm

Yard Waste
WWW: http://pigpen.cs.wisc.edu/simle/yard.html

Periodicals and Newsletters

BioCycle
E-mail: biocycle@aol.com

Bottle/Can Recycling Update
E-mail: resrecycle@aol.com

Composting News
E-mail: 71241.2763@compuserve.com

E Magazine
WWW: http://www.emagazine.com
E-mail: emagazine@prodigy.com

The Green Business Letter
WWW: http://www.enn.com/newsstand/gbl

Mill Trade Journal's Recycling Markets
E-mail: nvrecycle@aol.com

Modern Plastics
E-mail: modplas@ios.com

MSW Management
E-mail: MSW@rain.org

Paper Stock Report
E-mail: 71241.2763@compuserve.com

Plastics Recycling Update
E-mail: resrecycle@aol.com

Progress in Paper Recycling
E-mail: mahen@aol.com

Recycled Paper News
E-mail: apitzer@crosslink.net

Recycled Products Business Letter
E-mail: aso@aol.com

Recycling Canada
E-mail: SYDPUB@THE-FIX.SOS.ON.CA

Resource Recycling
E-mail: resrecycle@aol.com

Solid Waste Report
E-mail: 72110,1536@compuserve.com

Waste Age
WWW: http://www.envasns.org

Waste Reduction Tips
E-mail: aso@aol.com

World Wastes
WWW: http://www.intertec.com/wrldwst.htm

Worm Digest
WWW: http://www.applied3d.com/worm

Bulletin Board

Recycle Talk
WWW: http://grn.com/grn/mail/talk.htm

Recycle Talk is an online area reserved for questions and answers dealing with recycling issues ONLY. No commercial postings of any kind are allowed on this site, which makes it all the more valuable as a research tool.

Sources

Alta Vista search engine.
America Online, Inc.
American Plastics Council. *What Industry Is Doing*. Washington, DC: American Plastics Council, June 1996.
Colorado Department of Energy Conservation. *Recycle Colorado.* Denver, CO: Colorado Department of Energy Conservation, (Fall 1996).

Guttentag, Roger M. "Recycling in Cyberspace: The Searchers." *Resource Recycling* (April 1996): 76–77.

Recyclers World. World Wide Web: http://www.recycle.net/recycle/index.html

Resource Recycling. *The Directory of On-Line Resources in Recycling and Composting*. Portland, OR: Resource Recycling, 1996.

YAHOO! search engine.

Glossary

acid test A test that uses the acid phloroglucinol to test for the presence of groundwood in a paper grade. The acid reacts with the lignin in the groundwood and will turn the paper red/pink or purple.

alloy A substance composed of the mixture of two or more metals.

alumina Aluminum oxide. It is obtained from refining bauxite (it takes two pounds of bauxite to make one pound of alumina).

baler A machine in which recyclables are deposited, compacted, and compressed into wirebound bales to reduce their volume and facilitate their transport to a processing plant.

bauxite The ore from which aluminum is made. Aluminum oxide is obtained from bauxite, combined with cryolite salts in a solution, and subjected to direct current electricity to produce molten aluminum.

billet Aluminum that has been cast in log-shaped pieces, which is subsequently melted and forced through a die to make an extruded product (such as window frames).

bimetal can Any food or beverage can that has a steel body and an aluminum lid (considered 100 percent recyclable by the steel industry).

biodegradable Able to be broken down, by bacteria or other living organisms, into basic biological components (such as carbon dioxide and water) that can be assimilated back into the natural environment without causing any hazards. Most organic wastes, such as paper and food leftovers, are biodegradable under conditions allowing for the circulation of water, oxygen, and microorganisms.

bottle bill A law that requires a monetary deposit on beverage containers bought by consumers from retail establishments; the deposit may be refunded when the container is brought back to be recycled (see also **container deposit legislation**).

broker A person or group of persons acting as an agent or middleman between the sellers and buyers of recyclable materials.

buy-back center A place where persons bring their recyclables and exchange them for payment.

buy-back program A program that purchases recyclables from the public.

coated paper Paper coated with a material that improves its printing surface. The coating most frequently used for this purpose is clay.

combustion To burn; when used in the context of solid waste management, usually refers to the use of solid wastes, or components thereof (such as tires), as fuel (see also **incineration**).

commercial waste Waste that has originated in nonresidential places, such as offices, businesses, hotels, or warehouses.

commingle To mix recyclables with nonrecyclables. This trash will be collected and subsequently separated into its various recyclable components for further processing at a recycling center. Some curbside programs also allow commingling of recyclables only—e.g., aluminum and steel cans, glass and plastic containers, or any combination of these recyclables may all go in one container to be sorted at the recycling facility.

compost The stable end product of the decomposition of organic matter that results from the composting process (also called humus).

composting An aerobic (oxygen-dependent) degradation process by which plant and other organic wastes decompose under controlled conditions. A mass of biodegradable waste, in the presence of sufficient moisture and oxygen, undergoes "self-heating," a process by which microorganisms metabolize organic matter (their food source) and release energy in the form of heat as a by-product. This process is nothing more than an accelerated version of the breakdown of organic matter that occurs under natural conditions, such as on the forest floor.

container deposit legislation Laws that require that a monetary deposit be made when a certain type of container is bought. The theory behind these laws is that a required deposit will provide an incentive for consumers to return the containers to their place of purchase to collect the refund, thereby making recycling more probable (see also **bottle bill**).

contaminant When used in the recycling industry, this word refers to any substance that might be found mixed in with the product targeted for recycling. Most processors set strict limits as to the percentage of contaminants allowed in the materials they accept from sellers. Therefore, for example, glass would be a contaminant when mixed with aluminum cans that were going to be recycled into more cans, and an aluminum can would be a contaminant in a container of glass that was to be processed into recycled glass.

CPO Computer printout paper; fanfolded and usually with pastel-colored horizontal bars.

cullet Cleaned, crushed glass that is used to make glass products.

curbside collection Collection programs run by disposal/recycling companies that pick up recyclable materials, either source separated or commingled, from residential curbsides and deliver them to processing facilities.

decomposition A process by which something breaks down into its basic components or elements.

de-inking The mechanical and chemical process of removing ink from paper that has been printed on (especially newsprint).

depolymerization The breaking down of scrap plastics into their basic molecular parts, or polymers, part of the process of recycling plastic resins.

detinning A process whereby the tin is separated from the steel in a "tin" can; detinner companies buy steel cans and other tin mill products and, after detinning the recovered materials, sell the steel to steel mills and foundries and the tin to other markets.

discards Municipal solid waste that remains after recycling and composting have occurred. Discards, when not disposed of by littering or on-site disposal in rural areas, are generally landfilled or combusted.

diversion rate A measurement of the amount of material being diverted from the waste stream for recycling, compared with the total amount of waste that was discarded.

drop-off center A place where recyclable or compostable materials are brought and left for a recycling processor to collect. Typically, they are centrally located within a community to facilitate recycling.

end users Factories, mills, foundries, refineries, plants, and other businesses that use secondary or recyclable materials to manufacture new products.

ferrous A term describing iron-based products, including steel. Iron (chemical symbol Fe) gives steel its magnetic attraction. (This word is derived from the Latin word for iron: *ferrum.*)

flexible packaging Packaging such as bags, pouches, labels, liners, and wraps made of plastic film, paper, aluminum foil, or a combination thereof. Such packaging is compressible or collapsible, having no real shape of its own without the product is holds. It generally has a thickness of 10 millimeters or less.

flotation de-inking A process of removing ink from printed papers that uses a substance, such as clay, to bind with the ink and carry it off the paper fibers.

free sheet Paper that has been chemically treated to remove all lignin; paper that does not contain groundwood.

generation Refers to the amount, measured by weight or volume, of products and materials that enter the waste stream prior to recycling, composting, landfilling, or combustion.

groundwood Paper manufactured from pulp that has been acquired through a relatively inexpensive mechanical process and without the use of chemicals. It refers to a low-grade paper, usually used in its uncoated form in newspapers and in its coated form in magazines. Papers with groundwood will contain the chemical lignin and will react to the acid test by turning red/pink or purple.

head box The part of a papermaking machine that holds the pulp and distributes the pulp fibers onto the wire screen in a paper mill.

heavy metals Hazardous elements, such as cadmium, lead, and mercury, that can be found in the waste stream in such discarded items as light fixtures, batteries, coloring dyes, and inks.

HDPE High-density polyethylene.

high-density polyethylene A recyclable plastic often used for milk jugs, detergent bottles, and food containers, commonly known by its abbreviation, HDPE. Products made from this plastic are identified by the plastics identification code number 2.

high-grade papers Papers with longer wood fibers, generally considered more valuable than other paper grades. They include white ledger and computer printout papers.

hydropulper A machine, similar to a giant blender, used in the recycled paper industry to repulp used paper by churning it in hot water; the hydropulper reduces paper back to its fiber component.

incineration The burning of waste materials (also called combustion).

incinerator ash The material left as the combustion by-product of burned waste materials. This ash may also contain noncombustibles, such as metals.

inorganic waste Waste that is made up of materials that are neither plant nor animal, and therefore contain no carbon.

integrated solid waste management The practice of handling solid waste with several different prioritized approaches in order to recover the highest amount of energy possible from that waste. Such approaches usually include composting, recycling, source reduction, and waste-to-energy incineration, with landfilling used as the final option.

landfill A site where municipal solid waste is disposed of. Landfills are federally regulated by the Environmental Protection Agency under the authority of the Resource Conservation and Recovery Act (RCRA), specifically Subtitle D. Subtitle D ensures the environmental safety of a landfill from its inception to its closure, and even provides for regulation up to 30 years after closure. Leachate collection, groundwater monitoring, and the monitoring of methane gas are also required by Subtitle D. Landfills are becoming the last resort in an integrated solid waste management system, but they are considered the ultimate answer for wastes that cannot be recycled. Some landfills are specially engineered to recover the methane gas given off from decomposition. The gas is then burned to create steam, which in turn drives turbines. The turbines provide energy, which is subsequently sold back to electric utility companies.

leachate Liquid that has passed through solid waste and has pulled out various substances, including some that may be hazardous. The collection and treatment of leachate is a major concern at landfills, and Subtitle D (see **landfill**) mandates, in a ruling that took effect in 1993, that all operating landfills must have leachate collection systems conforming to EPA specifications.

lead-acid battery Generally known as an automobile battery; a type of battery where the main interior components consist of lead plates submerged in a sulphuric acid electrolyte solution.

lignin A naturally occurring chemical found in wood; a substance present in groundwood paper that will react with light, over time causing the paper to yellow and turn brittle.

magnetic separation A system in which a powerful magnet is used to separate ferrous metal objects from other materials in a mixed solid waste stream.

mandatory recycling Programs that legally require consumers to separate their garbage and remove specific items that must be collected for recycling.

manual separation The process of sorting and separating recyclables and/or compostable materials by hand from mixed solid waste.

materials recovery The practice of removing useful materials from the waste stream for recycling or reuse.

materials recovery facility (MRF) Usually refers to a place that accepts a variety of mixed wastes and separates the useful, recyclable materials from the rest either by human labor or mechanical extraction. MRF is pronounced "murf."

mechanical separation The process of sorting and separating recyclables and/or compostable materials from waste streams by mechanical means, such as magnets, screens, and blowers. Mechanical separation can also be used to sort commingled recyclables.

methane An explosive gas, odorless and colorless in its pure state, that is produced by municipal solid waste, or other organic matter, that is decomposing. Municipal solid waste landfills emit methane gas, and some landfills recover the gas for use in energy production.

mill broke Scrap paper produced from the paper manufacturing process; waste paper that has not been used by a consumer but has been produced, discarded, collected, and repulped within the papermaking facility in order to make new paper.

municipal solid waste (MSW) All nonhazardous waste produced in a city or town, including waste from households, businesses, and light industry. MSW may also contain mining waste and sewage sludge; definitions vary in different locales.

NIMBY An acronym for "Not in My Backyard," which refers to the general public's attitude toward the siting of solid waste facilities.

nonferrous A term describing metals that do not contain iron and thus have no magnetic properties.

nonrenewable Natural resources that cannot be replaced in their original form.

organic waste Waste matter that contains carbon, including wood, paper, yard wastes, and food wastes.

out-throws Papers of a grade different from the one being collected; papers manufactured or treated in a way that makes them unsuitable for use as the grade specified.

participation rate A measurement referring to the number of people participating in a given recycling program compared to the total number of possible participants.

pay-as-you-throw A solid waste management system in which residents of a community pay for garbage collection based on the amount of waste they generate, instead of paying a flat fee or property tax. Residents are thus given an economic incentive to reduce their waste through source reduction and recycling, since the amount of trash they put out on the curb is directly tied to the amount of money they owe on the bill they receive.

PET Polyethylene terephthalate.

phloroglucinol An acid solution, mostly hydrochloric in content, that reacts with the lignin in groundwood paper, turning the paper red/pink or purple. In the paper recycling industry, this solution is commonly called "groundwood tester."

plastics Manmade materials created from polymers. Polymers are large molecules containing mainly carbon and hydrogen, with smaller amounts of oxygen or nitrogen.

polyethylene A family of plastic resins created by the polymerization of ethylene gas, commonly grouped into two main categories: HDPE and PET.

polyethylene terephthalate A type of plastic resin, often used to make soft drink bottles. Commonly known by the abbreviation PET, it is identified by the plastics identification code number 1.

postconsumer recycling The reuse of materials that have been used by consumers, then collected from residential and commercial waste. This excludes materials that are the by-products of industrial manufacturing processes and have never reached the consumer (e.g., paper shreddings that have been cut off edges to make stationery).

preconsumer waste Waste produced in the manufacturing process; such waste has not been used and discarded by consumers.

pulp A solution of fibers and water used to manufacture paper.

recycling Any and all processes by which materials that would have otherwise become part of the solid waste stream are collected, sorted, and processed and returned to the consumer in a usable form as either raw materials or finished products.

resource recovery Refers to the beginning of the recycling process, where materials and energy are extracted from the waste stream (to be used in manufacturing new products), converted into another form of fuel, or used as an energy source.

scrap Any item or items that are no longer considered useful, e.g., appliances, cars, construction materials, and postconsumer steel cans. The term can also refer to new materials that are by-products of metals processing and manufacturing procedures. Steel scrap is recycled in steel mills and made into new products.

secondary material Another term for recyclable material; any material that has fulfilled its original use and can be collected and used again in another manner.

solid waste Garbage, refuse, sludge, and any other discarded solid materials. The term can be used to describe any such waste that originates from municipal, industrial, commercial, and agricultural operations.

solid waste management Those activities that concentrate on the systematic collection, source separation, storage, transportation, transfer, processing, treatment, or disposal of solid waste (see also **integrated solid waste management**).

source reduction The reduction of solid waste at its source, the consumer level. For example, environmentalists are encouraging shoppers to bring their own reusable cloth bags to grocery stores, thereby eliminating the use of either plastic or brown paper bags.

source separation The separation of recyclables (cans, glass, plastics, paper, etc.) at their source or point of origin, i.e., in the household, office, or school where the product is used and initially discarded.

steel A malleable alloy made of iron and carbon that is wholly recyclable.

steel can A rigid container made primarily or exclusively of steel; commonly known as the "tin" can, steel cans are often used to store food, beverages, and paint, and are 100 percent recyclable.

stickies Any non-water-soluble adhesive that will clog pulp preparation cleaning equipment or dissolve in the pulping process and later solidify in the finished paper product, causing spots and tears (e.g., self-adhesive address labels, self-sticking notes, and peel-and-press labels).

tin A metallic element that is often used as a coating on the inside of steel cans to stabilize the flavors of food contents.

tin can A term commonly used to describe a steel food can. When steel cans were first used to preserve foods, the tin coating on the inside was quite thick, but modern technology has reduced the tin to less than one-third of 1 percent of the weight of a steel can.

tipping fee The fee charged for unloading or dumping waste at a landfill, transfer station, or waste-to-energy facility, usually measured in dollars per ton.

transfer station A place where waste materials are taken from smaller collection vehicles, such as garbage trucks, and put into larger transporting units, such as semitrailers, for movement to a final disposal area, such as a landfill.

uncoated paper Paper that has not been coated with a substance to improve its printing surface.

virgin material Material that has never been used in a manufacturing process (e.g., virgin paper pulp is made from trees; virgin aluminum is made from ore; virgin resins are chemically created to make plastics).

virgin pulp Pulp made directly from plant fibers.

volume reduction A reduction in the amount of space that materials occupy, usually accomplished by mechanical, thermal, or biological processes.

waste stream An encompassing term referring to the waste material output of an area, location, or facility.

white goods Large appliances such as water heaters, refrigerators, and clothes washers and dryers. The term originated when white was the standard color for such items.

wire screen The part of a papermaking machine that picks the paper fibers up and out of the pulp and holds them in a thin, uniform layer while the excess moisture is removed.

Index